Vinicius A. Sikora de Souza

Extreme Precipitation in Western Amazonia

Vinicius A. Sikora de Souza

Extreme Precipitation in Western Amazonia

Rondônia - Brazil

ScienciaScripts

Publisher:
Sciencia Scripts
is a trademark of
Dodo Books Indian Ocean Ltd. and OmniScriptum S.R.L publishing group

120 High Road, East Finchley, London, N2 9ED, United Kingdom
Str. Armeneasca 28/1, office 1, Chisinau MD-2012, Republic of Moldova, Europe
Printed at: see last page
ISBN: 978-620-7-71556-5

SUMMARY

INTRODUCTION

Brazil has seen a significant growth in its urban population, especially since the 1960s. This phenomenon in recent decades has transformed Brazil into an essentially urban country (83 per cent urban population), generating an urban population with inadequate infrastructure (TUCCI, 1999; 2008).

The effects of this disorganised process have spread to the entire urban water resources system. Especially with regard to the planning and construction of hydraulic and hydrological works, in order to minimise socio-environmental impacts such as flooding through urban drainage planning (OLIVEIRA et al., 2011; TUCCI, 2007).

Knowledge of extreme hydrological events is a requirement in drainage, waterproofing and other engineering projects, whether in urban or rural areas, because it allows the designer to consider the risks involved in carrying out the work and associate them with the best alternative, from an economic point of view, without neglecting technical performance and safety issues. However, this data is incipient and limited in some locations.

Rainfall, among the hydrological elements, is the one that most interferes with human life, since it is the main input of water into the hydrological system, with other variants such as flow and infiltration being inextricably linked to its occurrence. Because of its wide influence on populated areas, whether positive or not, rain can be considered the main form of water supply for human and economic activities (ALMEIDA et al., 2011).

It is therefore of great importance to know and predict the characteristics of rainfall, with emphasis on elucidating its maximum intensity, the duration of this phenomenon and the period during which it may occur again.

In this sense, a widely used way of characterising extreme rainfall in a given location is the use of intensity-duration-frequency (IDF) curves. These consist of semi-empirical mathematical models that predict rainfall intensity through duration and temporal distribution. It should be noted that inferring extreme rainfall is possible because these events fit probabilistic distributions, allowing them to be modelled statistically.

In Brazil, rainfall volumes are essentially quantified by rain gauge stations in records known as daily rainfall and are the most accessible information, not only because of the size of the series, but also because of the density of the networks (HERNANDEZ, 2008). However, this data collection methodology causes an obstacle in the generation of IDF curves due to the unavailability of rainfall with shorter durations, which are fundamental in the process of modelling these curves.

In view of this problem, the 24-hour rainfall disaggregation method of the São Paulo State Environmental Sanitation Technology Company - CETESB (1979 apud TUCCI, 2009) presents itself as a solution, as it generates synthetic series with shorter duration intervals, using coefficients that transform 24-hour rainfall into shorter duration rainfall.

It should be noted that in the national territory, with the exception of the most developed locations, there are still regions that lack the development of these models for their territorial extensions, such as the state of Rondônia, located in the Western Amazon. This state is practically devoid of information on intense rainfall, forcing it to use information from nearby meteorological stations or stations with climatological characteristics similar to those of the location in which the project is being carried out as an alternative for carrying out hydraulic works projects.

This procedure, however, can lead to unreliable estimates, as this practice presents the risk of works being underestimated, causing mainly social problems, as well as being overestimated, increasing their cost.

OBJECTIVES

General objective

In view of the above, the general aim of this study was to estimate IDF functions based on rainfall data from rain gauge stations in the state of Rondônia, available on the website of the National Water Agency. This information can then be used in works in these locations.

Specific objectives

Specifically, the study has the following objectives:

a) Obtain the characteristics of heavy rainfall, such as intensity, duration and frequency;
b) Generate the areas of application for each IDF equation;
c) To analyse the efficiency of these equations by comparing the data estimated by them with the statistically modelled data;
d) Break down daily rainfall into 24-hour rainfall and shorter durations;
e) Check which of the seven statistical distributions best fits the distribution of extreme rainfall in the state;
f) To analyse the distribution of maximum rainfall values based on observations of rainfall data, verifying the spatial distribution of these events in the state's river basins and their possible influencing factors.

CHAPTER 1

THEORETICAL FRAMEWORK

1.1 HYDROLOGICAL CYCLE

Water is the most abundant inorganic substance on the surface of planet Earth, and this resource is estimated at $1.4.10^{15}$ m^3 (TEIXEIRA et al., 2009). According to Raghunath (2006), it is organised as follows: 97.2 per cent is salt water, which is mainly in the oceans, and only 2.8 per cent is available as fresh water. Of this 2.8 per cent of fresh water, 2.2 per cent is found on the planet's surface and 0.6 per cent as groundwater.

Raghunath (2006) also points out that of this 2.2 per cent of surface water, 2.15 per cent is in the solid phase in glaciers and ice caps and only 0.01 per cent is available in lakes and streams, the remaining 0.04 per cent is in other forms. Of the 0.6 per cent of groundwater stored, only 0.25 per cent can be economically extracted with current drilling technology.

According to Rossa (2006) all the water in the world makes up the hydrosphere, which is distributed in three main reservoirs: the oceans, the continents and the atmosphere, between which there is a constant circulation called the water cycle or hydrological cycle.

The water cycle is made up of different variables, which relate to each other through hydrological processes (PUYOL; VILLA, 2006). In this sense, the water cycle can be represented as a structure of volume in space, which is delimited by a boundary, whose internal components interact with each other and with adjacent systems (CHOW; MAIDMENT; MAYS, 1988).

This consideration is complemented by Tucci (2009), who defines the hydrological cycle as a global phenomenon of closed circulation of water between the earth's surface and the atmosphere, which is mainly driven by solar energy in association with gravity and the earth's rotation.

Tucci (2009) explains that the circulation of water in the Earth's system can occur in two directions: surface-atmosphere, in the form of vapour, which is considered the main element responsible for the continuous circulation of water around the globe; and atmosphere-surface, with water returning to the surface in liquid and solid phases through precipitation.

According to Collischonn and Tassi (2011), the hydrological cycle, which is diagrammed in Figure 1, can be illustrated as follows: water rises in the atmosphere

in the form of vapour due to heating by the sun and transpiration by plants. The water vapour then condenses to form clouds, and in specific circumstances the water contained in these clouds can precipitate, thus returning to the Earth's surface in the form of precipitation. Precipitation that reaches the surface can infiltrate into the ground or flow over the ground into a watercourse, where the infiltrated water moistens the soil, feeds aquifers and creates groundwater flow.

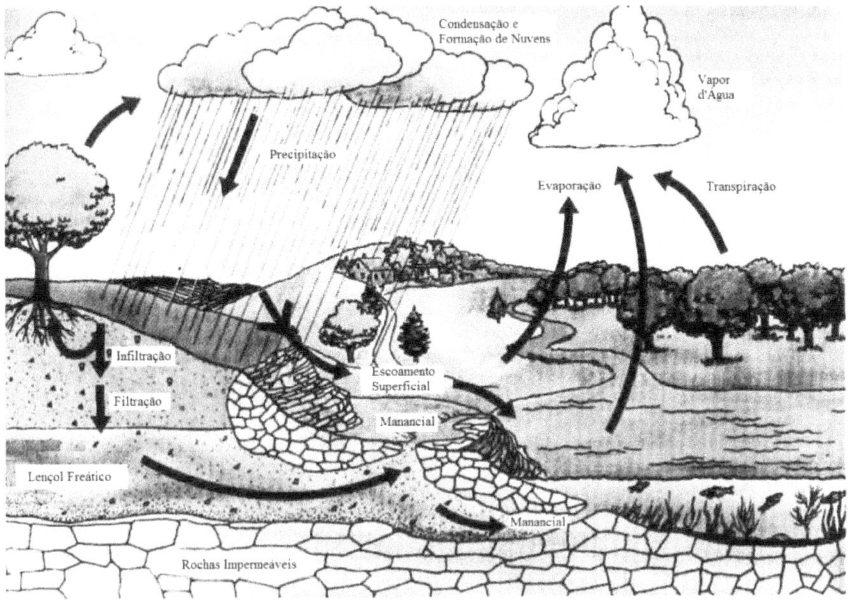

Figure 1 - Hydrological cycle.
Source: Adapted from Puyol and Villa (2006).

In this context, Feltrin (2009) conceptualises that the hydrological cycle can be understood as the system's water balance, the result of which will provide the water available in the system after various processes involving water flow. In this way, by applying the principle of conservation of mass and analysing the inflows and outflows of water in the system, the water cycle can be summarised using Equation 1 (ANIDO, 2002; RIGHETTO, 1998).

$$P = Ev + Q \pm \Delta S \tag{1}$$

Where:

P - Precipitation (mm);

Ev - Evapotranspiration (mm), and is the sum of the amounts of water evaporated and transpired by plants;

Q - Surface runoff (mm);

ΔS - Rate of water infiltrating or submerging the system (mm). This variable has both an addition and subtraction sign in the equation because it can be a source of both outflow and inflow of water;

however, in analyses of long series of the system, it can be disregarded because it is understood that the addition portion is of the same magnitude as the infiltration portion.

In this respect, precipitation is the main input variable in the water balance, so quantifying it, as well as knowing how it is distributed temporally and spatially, is essential in studies related to the need for irrigation, availability of water for domestic and industrial supply, soil erosion and flood control (DAMÉ; TEXEIRA; TERRA, 2008).

1.1.1 Precipitation

All water in solid or liquid form that comes from the atmosphere and reaches the earth's surface is called precipitation (TUCCI, 2009). As Zolet (2008) explains, precipitation includes all forms of moisture emanating from the atmosphere and deposited on the earth's surface, such as rain, hail, dew, fog, snow and frost.

According to Zolet (2008), rain, defined by Varejão-Silva (2005) as water droplets with a diameter of more than 0.5 cm, is the main element of the aforementioned that significantly interferes with engineering problems involving hydrological projects in Brazil.

In general terms, precipitation or rain is formed when the humid air in the lower layers of the atmosphere is heated, thus becoming less dense than the air in the upper layers, making it possible for it to rise adiabatically, expanding and cooling until it reaches condensation. Once these stages have taken place, the water vapour can condense into tiny droplets, thus creating clouds. These water droplets then undergo the processes of coalition and coalescence, which cause them to join together to form larger water droplets that will be able to overcome air resistance and precipitate (VILLELA; MATOS, 1975).

Rainfall can be classified according to the conditions that produce the vertical movement of air masses, which is a key factor in its formation, into cyclonic, orographic and convective rainfall (PUYOL; VILLA, 2006; REYNAUD, 2008).

Cyclonic rainfall occurs when two air masses with different temperatures, due to the unequal heating of the earth's surface, clash, thus generating a contact surface, also called a front. If these air masses move in such a way that the lower temperature air mass (cold front) "pushes" the warmer air mass (warm front), this rainfall is called frontal cyclonic rain. However, if the two air masses (cold and warm) are simultaneously drawn towards an area of low pressure, this phenomenon is called non-frontal cyclonic rain (RAGHUNATH, 2006). It should be noted that these rains are long-lasting, of low to moderate intensity and spread over large areas (VILLELA; MATOS, 1975).

As far as orographic rainfall is concerned, it is formed when hot, humid winds, usually from oceanic areas, come up against natural barriers, forcing these winds to rise and cool them adiabatically in the process, thus forming rainfall of low intensity, long duration and small spreading area (TUCCI, 2009).

The formation of convective rainfall occurs due to thermal convection, i.e. the sudden rise of humid air heated by the earth's surface which generates almost instant condensation. These events usually occur at the end of hot days. According to Tucci (2009), this phenomenon results in rainfall with small areas, high intensity and short duration.

1.1.1.1 Extreme rainfall events

Intense rainfall, as explained by Righetto (1998), is a meteorological phenomenon that causes flooding in drainage systems, where peak flows reach values close to the maximum capacity of these systems. Or, as Silva et al. (2003) explain, heavy rainfall, also known as extreme or maximum rainfall, is rainfall with a large amount of precipitation over a short period of time.

In general, heavy rainfall is capable of causing a large amount of surface runoff, which can cause major damage in agricultural areas, such as flooding of cultivated land, soil erosion, loss of nutrients, silting and pollution of water bodies (CECÍLIO et al., 2009). Cardoso, Ullmann and Bertol (1998) therefore emphasise the importance of knowing the characteristics of this rainfall in order to avoid the problems mentioned above and to help plan soil and water conservation practices, watershed management and the sizing of hydraulic structures in general.

In this context, the work of Beijo, Muniz and Castro Neto (2005) emphasises that hydrological data must be measured in the design of hydraulic works, thus seeking to reconcile the minimum cost for their implementation with the remote occurrence of risks of failure in a location.

Currently, the best solution for characterising and estimating rainfall is the use of Intensity-Duration-Frequency curves, which consist of semi-empirical mathematical models that predict rainfall intensity through duration and temporal distribution. Eltz, Reichert and Cassol (1992) state that frequency analysis is an important statistical technique in the study of rainfall, due to the great temporal and spatial variability of rainfall, which cannot be predicted on a purely deterministic basis.

According to Tucci (2009), these models are the main characteristics of rainfall. Therefore, they are estimates that aim to take into account the specific characteristics of rainfall in the location for which the model is made, by means of statistical analyses, because heavy rainfall adjusts to statistical distributions.

However, according to Cardoso, Ullmann and Bertol (1998), knowledge of rainfall characteristics is quite scarce in most of Brazil and, even in regions with a satisfactory density of rainfall stations, the available data is inadequate for immediate use, as it only shows intensities over periods of time greater than or equal to one day.

According to Pereira, Silveira and Silvino (2007), one of the probable causes of this scarcity is the fact that the country has a very large area, which makes it difficult to record such data. The lack of studies related to these records makes it difficult to draw up projects in the area of water resources in places far from Brazil's large urban centres.

The difficulty in generating models that describe the IDF relationship can be summed up in the availability of rainfall data and the low density of these records in Brazil; moreover, the methodology for obtaining them requires exhaustive work in tabulating, analysing and interpreting a large number of rain gauges (CECILIO; PRUSKI, 2003).

Various studies have been carried out proposing more efficient methods for the statistical adjustment of maximum rainfall data. However, there is a gap between theory and practice, which hinders the application of new techniques (DAUD et al., 2002).

Some methodologies have been developed in Brazil to obtain intense rainfall of shorter duration from rainfall data. These methodologies use coefficients to transform one-day rainfall into rainfall of shorter duration.

Among these methodologies, the 24-hour rainfall disaggregation method of the São Paulo State Environmental Sanitation Technology Company - CETESB (1979 apud TUCCI, 2009) stands out. The use of this method is very efficient, as it has been applied in several studies, including Cardoso, Ullmann and Bertol (1998); Oliveira et al. (2000); Oliveira et al., 2011; and Pereira, Silveira and Silvino (2007).

Furthermore, the use of this methodology was indicated in the study by Darné, Teixeira and Terra (2008), because when they analysed the precipitation series for Pelotas - RS, it corresponded more reliably to the data.

Back (2009) also points out that this method has the advantage of being simplified, as well as providing satisfactory results with great similarity for different locations for which the coefficients are generated.

Thus, by using the 24-hour rainfall disaggregation technique of CETESB (1979 apud TUCCI, 2009), it is possible to solve the problems pointed out by Silva et al. (1999) regarding the short period of observations available for estimating the parameters of the rainfall intensity-duration-frequency equation, which according to Silva et al. (2003) is the main way of characterising heavy rainfall. Tucci (2009) also emphasises that this method is highly applicable, given the large number of rain gauges with long series, spread over almost the entire national territory.

1.1.1.2 Average rainfall in an area

Average rainfall is understood to be all liquid water with uniform height characteristics over a given area of influence. In this sense, Sanchez (1986), quoted in Tucci (2009), emphasises, however, that this ideology is an abstraction, because in real terms the phenomenon obeys variable temporal and spatial distributions.

Tucci (2009) reports on the existence of various methods for determining average rainfall in a given area, where these make use of weightings of the data measured by rain gauge stations, with the arithmetic mean, Thiessen polygons and isohyets being the most commonly used methods.

The arithmetic mean method is based on the assumption that all the rain gauges have the same coverage in the area in which they are located, i.e. that they exert their influence on the average rainfall in a similar way. In this context, according to Puyol and Villa (2006), this variable can be calculated using Equation 2.

$$P_m = \frac{1}{n}\sum_{i=n}^{n} Pi \tag{2}$$

Where:

P_m - Average rainfall (mm) over a given area;

Pi - Precipitation in the nth rain gauge (mm);

n - Total number of rain gauges used in the calculation.

The Thiessen method considers that each rainfall station will have a sub-area of influence within the area in which it is located, so this methodology predicts that these stations will be arranged non-uniformly in space. These sub-areas of influence are determined using Voronoi diagrams, which form polygons using the medians of the lines linking two adjacent stations (PRUSKI et al., 2004).

This methodology can sometimes be unfeasible in some situations because it does not take into account the influence of the terrain, and in order to present significant results the terrain must be slightly hilly and the distances between the rain gauges not very long (TUCCI, 2009). Therefore, the average rainfall for a given area is obtained using Equation 3 (PRUSKI et al., 2004).

$$P_m = \frac{\sum_{i=1}^{n}(P_i.A_i)}{\sum_{i=1}^{n} A_i} \tag{3}$$

In which:

Ai - area of influence (m^2) of each rainfall station.

Isohyets are defined as curves of the same rainfall. They are produced by interpolating data from rain gauge stations and then adjusting them according to the relief of the area in which they are located. Tucci (2009) describes that the calculation of average rainfall (P_m) using this analysis methodology is carried out using Equation 4.

$$P_m = \frac{1}{A}\left(\frac{P_i + P_{i+1}}{2}\right) \sum A_{i,i+1} \tag{4}$$

Where:

A - is the total area (m)$;^2$

P_{i+1} - rainfall (mm) of the nth adjacent rain gauge.

1.2 RIVER BASIN

A hydrographic basin is defined as a natural catchment area for precipitation water that converges the runoff into a single outlet (CARDOSO et al., 2006). Vilela and Matos (1975) emphasise that the term watershed designates a certain area where precipitation events are collected and transported to its natural drainage system.

Complementing the other authors mentioned, Prioste (2007) reveals that a river basin is a physiographic unit that collects precipitation and acts as a reservoir of water and sediment, flowing into a single river section, which is called the outflow. The author also emphasises that river basins are bounded by topographic dividers or watersheds, which are the crests of elevations in the terrain that separate the drainage of precipitation between two adjacent basins.

Rodrigues and Adami (2011) also reveal that a river basin is a system comprising a volume of materials, predominantly solids and liquids, close to the earth's surface, delimited internally and externally by all the processes that, from the supply of water by the atmosphere, interfere with the flow of matter and energy in a river or a network of river channels.

Argento and Cruz (1996) emphasise that the delimitation of river basins is based on contour lines, drawing a water dividing line that connects the highest points - tops - of the region around the drainage area in question. Thus, the study of a watershed must begin with the Topographic Map, because as well as enabling delimitation, it provides basic location elements, such as reference elements, systematisation elements and proportion elements (CASTRO, 2000).

In view of these facts, the first step in analysing a basin is to identify the lowest point on the map, which is known as its outlet. Once the lowest point in the basin has been identified, certain parameters must be calculated to help describe and quantify the characteristics of the basin. These analyses will provide useful information for making decisions on how to manage the basin, as well as simply describing it (DILL, 2007).

1.3 HYDROSTATISTICS FOR EXTREME EVENTS

In studies of the water cycle, especially in relation to hydrological aspects, statistical analyses are an important tool for studies and decision-making on this subject (TUCCI, 2009).

In this respect, this section will take a cursory look at some concepts of statistical distributions with a focus on extreme events. It is worth noting that Leotti, Birck and Ribold (2005) report that these theoretical distribution models seek to represent the behaviour of a given event according to the frequency of its occurrence.

Therefore, these distributions are actually probability distributions, where for an event we have an associated probability of occurrence. In other words, we can infer how likely it is that an event will occur again.

1.3.1 Normal Distribution

The Normal distribution, also known as the Gauss distribution, is used to describe the behaviour of a random variable that fluctuates symmetrically around a central value, so this distribution is appropriate for modelling variables that result from the sum of a large number of other independent variables (NAGHETTINI; PINTO, 2007).Furthermore, according to Silvino et al. (2007), the Normal distribution is at the origin of all the theoretical formulation about the construction of confidence intervals, statistical hypothesis tests, as well as regression and correlation theory. However, the author points out that climatological data may not follow this statistical distribution due to its heterogeneity.

1.3.2 Log-Normal Distribution

The Log-Normal distribution follows the same assumptions as the Normal distribution (SILVINO et al., 2007). However, Tucci (2009) points out that when compared, the Log-Normal distribution gives better results in modelling hydrological processes, as shown by Catalunha et al. (2002), who state that in the study by Huf and Neili (1959), the Log-Normal distribution gave satisfactory results when compared with other methods for analysing rainfall frequency.

1.3.3 Exponential distribution

The Exponential distribution is a distribution of extremes. It is an asymmetric model that has numerous other applications in various areas of human knowledge and, in particular, to hydrological variables (NAGHETTINI; PINTO, 2007). As Lanna (1993, apud SILVINO et al., 2007) points out, the exponential distribution approximates the shape of the frequency distribution of certain maximum annual hydrological events.

1.3.4 Gama Distribution

The Gamma distribution is defined as a continuous and asymmetrical distribution. Because it is directly related to theories of reliability and random counting, it is often used to represent physical events, such as analysing the duration and intensity of storms (PEREIRA, 2010).

Tucci (2009) reports that because this distribution has a long upper tail, it is representative of extreme hydrological events where asymmetry is present.

1.3. 5 Gumbel distribution

The Gumbel distribution, also known as the double-exponential distribution or the Fisher-Tippet Type I distribution, is part of the extreme values of Type I, which has its tails as a function of the Euller constant.

It should also be noted that the asymptotic form of this distribution is widely used in analysing the frequency of hydrological events (BACK, 2001; NAGHETTINI; PINTO, 2007; TUCCI, 2009).

1.3. 6 Weibull distribution

The Weibull distribution is a discrete distribution based on shape and scale parameters. This distribution is similar to the gamma distribution (PEREIRA, 2010).

It should be noted that it has been used in hydrological modelling for extreme events, such as the studies by Barbosa et al. (2005), Catalunha et al. (2002), Lima (2005), Queiroz et al. (2010) and Sá (2011).

1.3. 7 Logistical distribution

The Logistic distribution is a continuous distribution whose purpose is to model the occurrence of relative frequencies. It resembles the normal distribution in shape, but has denser tails due to greater kurtosis.

The name of this distribution derives from the distribution function, which is an instance of a member of the logistic functions. This distribution was used in the study by Pinheiro and Naghettini (1998) to regionally analyse the frequency and temporal distribution of storms in some regions of Belo Horizonte - MG.

CHAPTER 2

MATERIAL AND METHODS

2.1 STUDY AREA

2.1.1 Location and General Characteristics

The state of Rondônia (FIGURE 2) is located in the Western Amazon between parallels 7° 58' and 13° 43' South Latitude and meridians 59° 50' and 66° 48' West Longitude. Rondônia is made up of 52 municipalities, the most populous of which is the state capital, Porto Velho, which has 428,527 inhabitants (IBGE, 2010).

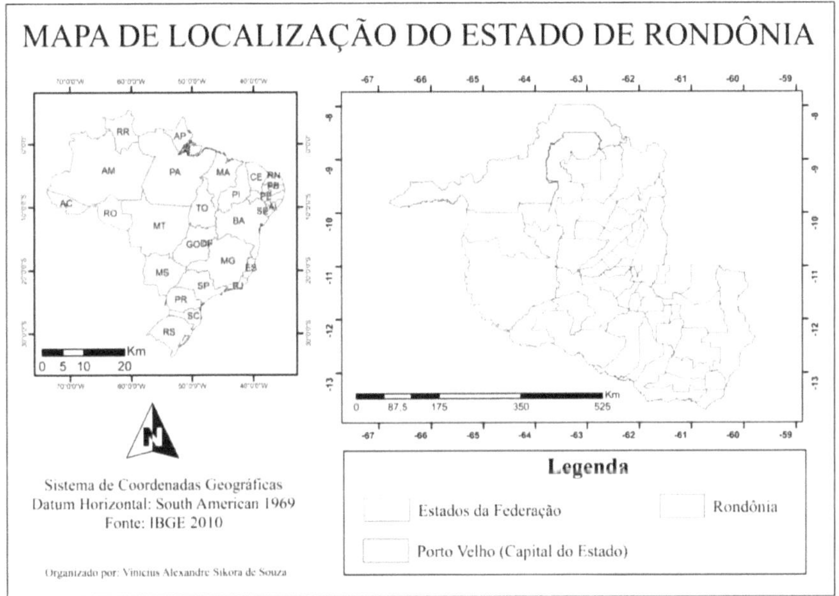

Figure 2 - Location of the state of Rondônia.

The state of Rondônia has altitudes ranging from 70 to 600 metres, and its main relief units are: Amazonian Plains; Depressions (Southern Amazon, Guaporé and Solimões); Plateaus (Southern Amazon and Parecis); and the Guaporé Pantanal. The types of vegetation found in the region are Open Ombrophilous Forests, Dense Ombrophilous Forests (Amazon Rainforest), Semideciduous Seasonal Forests, Savannas and Alluvial Vegetation, with Dense Forests predominating in the northwest of the state (BRASIL, 1978).

2.1. 2Climate

According to the Köppen classification, the climate of the state of Rondônia is characterised as AW (tropical-hot and humid), with a climatological average air temperature during the coldest month of more than 18 °C (megathermal) and a well-defined dry period during the winter season, when there is a moderate water deficit in the state with rainfall rates of less than 50 mm/month (RONDÔNIA, 2009).

Rainfall in Rondônia is monsoonal and therefore only occurs during the time of year when the continent is heated above ocean surface temperature levels and becomes a convection centre, where there is an ample supply of moisture for condensation (BUTT; OLIVEIRA; COSTA, 2011).

Annual rainfall in the central region of the state averages over 2,000 mm, with an average relative humidity of 82 per cent and an average annual temperature ranging from 24 °C in the rainy season to 25 °C in the dry season (AGUIAR et al., 2006; WEBLER; AGUIAR; AGUIAR, 2007).

According to the State Secretary for Environmental Development (SEDAM) (2009), this region is not greatly influenced by continentality, i.e. greater or lesser distance from the sea.

2.1. 3Hydrography

The hydrographic network of the state of Rondônia is made up of the Madeira River and its tributaries, as well as floodplain lakes that interact with the rivers. According to Zuffo and Silva (2003), this is due to the layout of the Parecis and Pacaás Novos plateaus, which run predominantly from southeast to west, forming the great surface drainage divide at state level, with a predominantly radial-dentritic pattern.

Figure 3 shows the state's main rivers, with emphasis on the Madeira, Machado (or Ji-Paraná), Mamoré, Guaporé and Jamari: Madeira, Machado (or Ji-Paraná), Mamoré, Guaporé and Jamari.

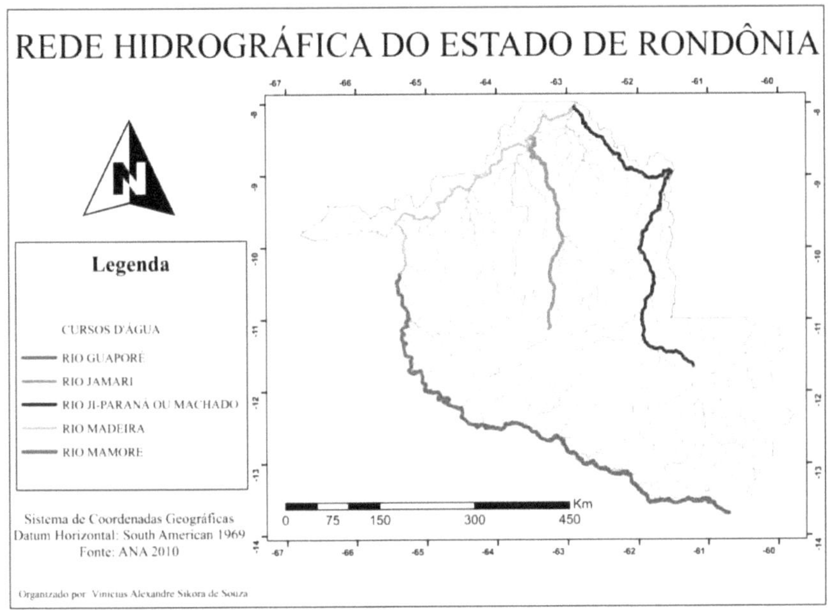

Figura 3 - Rondônia's hydrographic network.

The state of Rondônia was divided into seven river basins, according to a study carried out in 1999, the aim of which was to subsidise the State Law on Water Resources and which involved the Regional Council of Engineering, Agronomy and Architecture (CREA/RO), the Government of Rondônia and the Ministry of the Environment (MMA) (SILVA; ZUFFO, 2002). Table 1 summarises some of their characteristics and Figure 4 shows their locations.

Table 1 - Hydrographic basins in the state of Rondônia.

Code	Name	Area (km)^2
01	Guaporé River Basin	59.339,380
02	Mamoré River Basin	22.790,663
03	Abunã River Basin	4.792,210
04	Madeira River Basin	31.422,152
05	Jamari River Basin	29.102,708
06	Machado River Basin	80.630,566
07	Roosevelt River Basin	15.538,192

Source: Silva and Zuffo (2002).

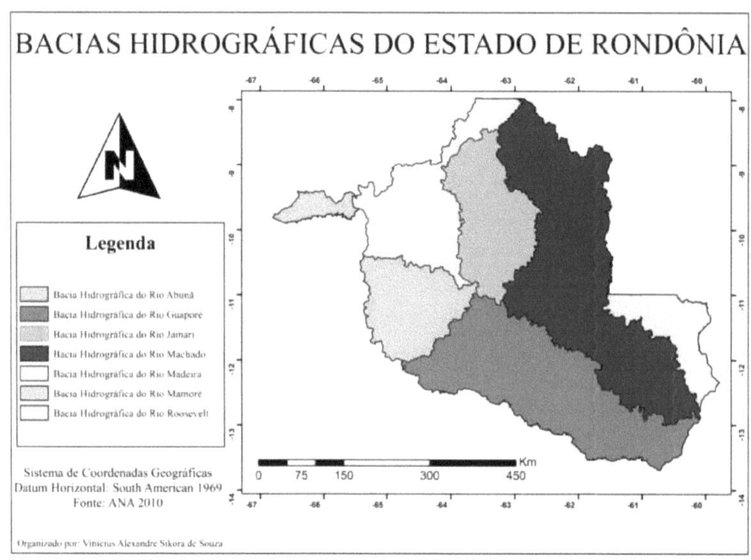

Figura 4 - Hydrographic basins in the state of Rondônia.

2.2 . ANALYSING THE DATA

2.2.1 Data Acquisition

The rainfall data used in this study was obtained from the database provided by the National Water Agency (ANA). It should be noted that information from this database has been used in several similar studies, including: Arantes et al. (2009); Butt, Oliveira and Costa (2011); Fietz and Comunell (2006); and Oliveira et al. (2010).

This agency keeps records of accumulated rainfall data for one-day periods for 93 rainfall stations in the state of Rondônia, but only 41 of them had a set of data with a time series of 10 years or more. This characteristic was explained in the study by CETEB (1979, apud FIETZ; COMUNELL, 2006) as a requirement for the IDF equations, which are the subject of this study. Table 2 gives more detailed information on the stations used in the study and Figure 5 shows their locations.

Table 2 - Data from the rainfall stations sampled.

Station	Code	Series (years)	Effective months	Municipality	Responsible	Location Latitude	Longitude
1	1360002	1983 - 2010	333	Pimenteiras do Oeste	ANA	13°28' 47" S	61° 02' 47"O
2	1360001	1983 - 2010	336	Cherry trees	ANA	13° 11'48" S	60° 49' 24" O
3	1360000	1983 - 2010	334	Colorado do Oeste	ANA	13° 06' 51" S	60° 32' 54" O
4	1264000	1983 - 2010	329	Costa Marques	ANA	12° 25' 37" S	64° 25'21"

5	1262001	1999 - 2010	139	Alta Floresta do Oeste	ANA	12° 36' 05" S	62° 10' 42" O
6	1262000	1980 - 2010	352	Costa Marques	ANA	12°51'05" S	62° 53' 57" O
7	1261001	1999 - 2010	130	Parecis	ANA	12° 12' 33"S	61° 37' 43" O
8	1261000	1983 - 2006	260	Chunpinguia	ANA	12° 29' 16" S	61° 02' 47" O
9	1260001	1967 - 2003	337	Vilhena	DEPV	12° 42' 00" S	60° 05' 00" O
10	1164001	1983 - 2002	229	Guajará-Mirim	ANA	11°22' 00" S	64°53' 11" O
11	1164000	1976 - 1989	167	Guajará-Mirim	ANA	11°04' 00" S	64° 05' 00" O
12	1161003	1999 - 2010	131	Minister Andreazza	ANA	11° 11'49" S	61°31'41" O
13	1161002	1983 - 2010	315	Rolim de Moura	ANA	11°44'59" S	61° 46" 35' O
14	1161001	1980 - 2010	372	Pimenta Bueno	ANA	11°41'01" S	61° 11'32" O
15	1161000	1977 - 2010	395	Cacoal	ANA	11°26' 27" S	61° 29' 02" O
16	1160002	1982 - 2010	334	Pimenta Bueno	ANA	11°44'56" S	60° 52'04"O
17	1160000	1977 - 2010	393	Pimenta Bueno	ANA	12° 00' 55" S	60°51' 18" O
18	1065002	1972 - 2010	451	Guajará-Mirim	ANA	10° 47' 33" S	65° 20' 52" O
19	1063001	1980 - 2010	354	Ariquemes	ANA	09° 45' 39" S	63° 17' 15" O
20	1063000	1978 - 2010	358	Ariquemes	ANA	10°30' 18" S	63° 38' 46" O
21	1062004	1986 - 2010	285	Theobroma	ANA	10° 14' 11" S	62° 20' 45" O
22	1062003	1983 - 2010	324	Mirante da Serra	ANA	11°00' 13" S	62° 39' 22" O
23	1062002	1978 - 2010	383	Jaru	ANA	10° 14' 11" S	62° 37' 38" O
24	1002001	1977 - 2010	399	Jaru	ANA	10° 20' 45" S	62° 27' 50" O
25	1061003	1987 - 2010	285	Ouro Preto do Oeste	ANA	10°31'01" S	62° 00' 05" O
26	1061001	1975 - 1997	262	Ji-Paraná	ANA	10° 50' 58" S	61° 55'50" O
27	966001	1982 - 2010	316	Old harbour	ANA	09°41'25" S	65° 59' 35" O
28	966000	1977 - 2010	404	Old harbour	ANA	09° 45' 20" S	66° 36' 42" O
29	965001	1976 - 2010	377	Old harbour	ANA	09° 42' 11" S	65°21'53" O
30	964001	1978 - 2002	290	Old harbour	ANA	09° 30' 59" S	64° 48' 44" O
31	963009	1997 - 2010	162	Ariquemes	ANA	09° 28' 00" S	63° 15'00" O
32	963006	1983 - 2001	223	Ariquemes	ANA	09° 53'14"S	62° 59' 17" O

33	963004	1980 - 2010	366	Ariquemes	ANA	09° 53'14"S	62° 59' 16" O
34	963001	1977 - 2010	387	Old harbour	ANA	09° 15'38" S	63° 09' 43" O
35	963000	1975 - 2010	420	Ariquemes	ANA	09° 55' 54" S	63° 03 '25" O
36	962001	1981 - 2007	308	Old harbour	ANA	09° 10' 45" S	62°57' 11" O
37	962000	1978 - 2010	378	Machadinho do Oeste	ANA	09°35' 11" S	62° 23' 38" O
38	961003	1987 - 2009	273	Machadinho do Oeste	ANA	09° 40' 53" S	61° 58' 44" O
39	863003	1975 - 2002	297	Candeias do Jamari	ANA	08° 45'35" S	63°27' 45" O
40	863000	1961 - 2007	564	Old harbour	INMET	08° 46' 00" S	63° 55'00"O
41	862000	1977 - 2010	385	Machadinho do Oeste	ANA	08° 56' 00" S	62° 03' 14" O

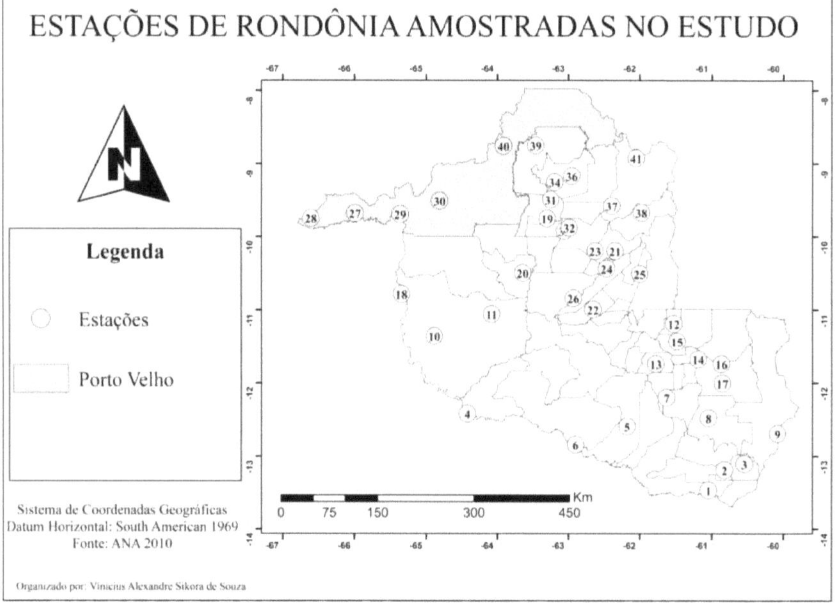

Figure 5 - Location of the rainfall stations in the state of Rondônia used in this study.

2.2.2 Adherence test

In order to generate the IDF equations for the state of Rondônia, it was necessary to check which statistical distribution best fits the maximum annual rainfall events recorded in the data collected, using an adherence test.

The distributions tested, for reasons described in section 1.3 of this book, were the Normal, Log-Normal, Exponential, Gamma, Gumbel, Weibull and Logistic distributions.

Among the adherence tests indicated by Naghettini and Pinto (2007), we opted for the Kolmogorov-Smimov (KS) test, which according to the aforementioned authors is a non-parametric test based on the maximum difference between the empirical and theoretical cumulative probability functions of continuous random variables, which is conservative in terms of the magnitude of the type I error.

This test is also highlighted by Leotti, Birck and Ribold (2005) as belonging to the supreme class of statistics based on the empirical distribution function, as it works with the greatest difference between the empirical and hypothetical distributions. Catalunha et al. (2002) also explain that KS is widely used to analyse the adherence of distributions in climate studies.

Thus, the Kolmogorov-Smimov test for the distributions analysed, carried out using the Weibull++ 7® software (RELIASOFT, 2011), was first carried out by conjugating the null hypothesis (Ho), that the empirical distribution is the same as the distribution of the sample values, and the alternative hypothesis (Hi), where the assertion of the null hypothesis would not be true, i.e. the distribution of the data analysed would not correspond to the statistical distribution analysed.The decision criterion was a comparison of the parameters D_{ca} ic and D_{cr} it, where D_{ca} ic > D_{cr} it rejects the Ho hypothesis in favour of Hi. In the adherence tests, a significance level (a) of 0.005 was used to obtain a reduced value for the type I error, i.e. minimising the chances to approximately 100% of discarding the null hypothesis if it is true (BORGES; FERREIRA, 2003).

2.2.3 Construction of IDF equations

In order to estimate the IDF equations, the maximum "one day" rainfall for each year was first obtained for the historical series of each rainfall station analysed, thus constituting the series of maximum annual rainfall. The data was then organised in descending order and the arithmetic mean and standard deviation of the sample were calculated. This procedure made it possible to statistically analyse the probability and return period of heavy rainfall using the Gumbel distribution method, the use of which is explained in section 3.2 of this study.The reduced Gumbel variable (y) was obtained using Equation 5, as recommended by Gumbel (2004).

$$y = \frac{s_y}{s_x}\left[x_i - \left(x_m - s_x \frac{y_m}{s_v}\right)\right] \tag{5}$$

Where:

s_x - Standard deviation of the series;

x_I - Value of a sample element;

x_m - Average of the sample of the finite annual series of n values;

s_y - Standard deviation, tabulated value;

y_m - Average of the reduced variable (y), tabulated according to the number of data points in the sample.

The return period (Tr), defined as the average interval, in years, in which any given rainfall value is equalled or exceeded at least once, is estimated by Equation 6. This expression is a function of the base of the Neperian logarithms (e).

$$Tr = \frac{1}{1-e^{-e^{-y}}} \tag{6}$$

After this procedure, the data was plotted on a graph with the same characteristics as the log-probabilistic paper, also known as the Gumbel paper, i.e. the points corresponding to the maximum rainfall heights (p) were on the ordinate, on an arithmetic scale, and the corresponding return period, in years, on the abscissa, on a log-probabilistic scale.After this action, a straight line was fitted to the data on this graph, which covered the range of data analysed, as they had a coefficient of fit (r^2) varying from 89% to approximately 100%, as can be seen in Table 3.

In this way, it was possible to estimate the maximum "one-day" rainfall for various return periods, and even extrapolate information for return periods longer than those contained in the data interval.

Table 3 - Adjustment coefficients of the equations estimated for the data from the stations analysed.

Code	Station	r^2	Code	Station	r^2	Code	Station	r^2
1360002	1	0,901	1161000	15	0,938	965001	29	0,988
1360001	2	0,967	1160002	16	0,976	964001	30	0,961
1360000	3	0,962	1160000	17	0,951	963009	31	0,896
1264000	4	0,970	1065002	18	0,982	963006	32	0,894
1262001	5	0,968	1063001	19	0,974	963004	33	0,967
1262000	6	0,922	1063000	20	0,991	963001	34	0,960
1261001	7	0,984	1062004	21	0,980	963000	35	0,971
1261000	8	0,987	1062003	22	0,896	962001	36	0,962
1260001	9	0,978	1062002	23	0,941	962000	37	0,970
1164001	10	0,929	1062001	24	0,964	961003	38	0,954
1164000	11	0,950	1061003	25	0,918	863003	39	0,963
1161003	12	0,926	1061001	26	0,954	863000	40	0,995
1161002	13	0,950	966001	27	0,931	862000	41	0,919
1161001	14	0,982	966000	28	0,943	-	-	-

After obtaining the rainfall heights for the periods from 2 to 100 years, the probable maximum average intensities for all rainfall durations from 5 minutes to 24 hours were estimated by disaggregating the daily rainfall, using the national average ratio quotients (TABLE 4) obtained by CETESB (1979), as explained in Tucci (2009).

Table 4 - Quotients at national level of the relationships between durations.

Relationship	Quotient
5min/30min	0,34

10min/30min	0,54
15min/30min	0,70
20min/30min	0,81
25min/30min	0,91
30min/1h	0,74
1h/24h	0,42
6am/24pm	0,72
8am/24pm	0,78
10am/24pm	0,82
12/24h	0,85
24h/day	1,14

Source: Tucci (2009)

After obtaining the maximum heights for the desired periods and durations, the IDF equation was generated for the areas of influence of each rainfall station in Rondônia by establishing the constants - K, a, b and c - using the least squares method. For the general IDF equation, Equation 7, which according to Villela and Mattos (1975) is the most widely used mathematical model for expressing the rainfall IDF relationship. The software LAB Fit® (SILVA et al., 2004) was used, which its creators claim is an effective tool for curve fitting.

$$i_m = \frac{K.Tr^a}{(t+b)^c} \tag{7}$$

Where:

i_m - Maximum average rainfall intensity, mm/h;

t - Duration (min);

Tr - Return period (year);

K, a, b, c - Parameters relating to the location.

2.2.4 Analysing the Efficiency of IDF Curves

To check the efficiency of the IDF equation proposed in this study, the regression coefficient (r^2) for fitting the function to the points was used and the Wilcoxon-Mann-Whitney hypothesis test, available in MINITAB® Statistical Software, 16.0 demo (MINITAB, 2011) was used to check whether the data measured and modelled according to the Gumbel distribution differ statistically from the data estimated by the IDF function. The null hypothesis (Ho) was that these data are the same, i.e. that the average difference between the modelled data and the estimated data is equal to magnitude 0, and the alternative hypothesis (Hi) was that they differ from each other, i.e. that the average difference is not equal to 0.

The decision criterion was based on the interval constructed for these differences with a confidence level of 99.5 per cent, this magnitude being established for the same reasons as the choice of a for the KS test, so if the value 0 is contained in this interval the null hypothesis is accepted, as the mean is statistically equal to 0, otherwise the alternative hypothesis would be accepted as the previous assertion becomes untrue.

It should be noted that Prazeres Filho, Viola and Borges (2010) report that the Wilcoxon-Mann-Whitney test is a non-parametric test, which does not require the original distribution of the data and is therefore called a distribution-free test. It is indicated for testing whether two samples are identical or not, and is also used when the variables studied are measured on a scale at least at ordinal level, and is an alternative to the paired t-test.With regard to checking the residuals produced by the differences in the magnitude of the extreme rainfall data disaggregated and modelled by the Gumbel distribution (XM) and the data estimated by the IDF equations drawn up in this study (XE), analysis was carried out of the average standard error (EPM), Equation 8; average normalised error (ENM), Equation 9; and average multiplicative error (EMM), Equation 10. All these analyses are explained in the work by Moog and Jirka (1998).

$$EPM = \sqrt{\left[\sum_{i=1}^{n} \frac{(x_E - x_M)_i^2}{N}\right]} \tag{8}$$

$$ENM = \frac{100\%}{N}\sum_{i=1}^{n}\left(\frac{x_E - x_M}{x_M}\right)_i \tag{9}$$

$$EMM = e^{\left[\frac{\sum_{i=1}^{n}\left|ln(x_E/x_M)_i\right|}{n}\right]} \tag{10}$$

2.3 DETERMINING THE AREAS OF INFLUENCE OF IDF CURVES

Based on the premise that the municipalities in the state of Rondônia have large territorial extensions, this study opted to indicate the use of the IDF equations generated not for each municipality specifically, but for the area of influence of the equation.

The theory used to construct these areas of influence was the Voronoi diagram, with each area representing a cell in this diagram, whose characteristic is the intersection of all the half-spaces defined by bisectors, i.e. the medians of the lines connecting two adjacent rainfall stations, which were used in the equation (BOOTS, 1986).

It should be noted that this theory for constructing areas of influence is widely used in rainfall analyses, such as the Thiessen polygons, which are described in section 3.1.1.2 of this book.

The areas were created using Quantum Gis® software (SHERMAN et al., 2011), where the geographical coordinates of the rainfall stations sampled were first converted from a table into a *shape file* (.shp), which was then used to create the polygons of influence by selecting the "Voronoi diagram" option in the programme's menu bar. After this procedure, the diagram generated was intersected with the vector file for the state of Rondônia, making it possible to generate and identify the areas of influence of the state's rainfall stations.

2.4 MEASURING THE MAXIMUM AVERAGE RAINFALL EVENTS FOR THE RIVER BASINS

The techniques used to determine the average extreme rainfall for each of the seven river basins in the state of Rondônia were the arithmetic mean, Thiessen polygons and isohyets.

The arithmetic mean was calculated by first obtaining the intensity data for each return period and duration from the rain gauge stations in the basin, and then using this data in Equation 2 to replace the rainfall height variable, so that the result corresponds to the maximum average intensity for the basin.

To determine the maximum rainfall using the Thiessen polygon method, the geometric element corresponding to the intensity of each rainfall station for a specific duration and return time was first generated, and the procedures for this generation were similar to the creation of areas of influence described in section 1.3. Once these data were available, they were applied to Equation 3, in place of the precipitation height.

The isohyettes were generated in Quantum Gis® software (SHERMAN et al., 2011) by first importing the data table of extreme rainfall intensities in their specific return periods and durations, including the geographical locations of the stations, using the "delimited text" *plugin.*

To limit the analysis to the area covered by the rain gauges, a simple minimum convex shape was created by selecting the "convex shape(s)" item in the "Geoprocessing Tools" sub-menu of the "Vector" menu. In order to expand the analysis to the entire state, a "Bufer" of the shape created was generated, this option also being located in the "Geoprocessing Tools" sub-menu.

After these steps, the resulting geometry was intersected with the *shape file of* the Rondônia contour. After creating the isohelines, the geometric shape created in the previous steps was imported, along with the vector of station points, into an extension called "GRASS" in the software itself, using the "v.in.ogr.qgis" module.

Once these actions had been carried out, *the* intersected geometry was converted into a raster using the "v.to.rast" command, so that it could be used as a mask. Subsequently, the "r.mask" module was used to force the newly generated *raster to* approximate the vector of rainfall stations.

"v.surf.rst" was used to create the interpolated grids of extreme rainfall intensities, selecting as an attribute field to interpolate the column containing the rainfall intensity with the Tr and t of interest. As a final procedure for generating the grids, they were converted into curves using the "r.contour" command.

Once the isopleths were available, the relevant data was extracted and applied to Equation 4 to calculate the average, stressing that once again the precipitation height was replaced by the intensity.

It should be noted that after obtaining the average intensity data for each of Rondônia's hydrographic basins using each of the methods described above, they were submitted to the Kruskal-Wallis (K-W) non-parametric statistical test at 0.001, since this is the lowest level of confidence that would result in the lowest type I error that the software works with, which was carried out using Graphpad Prism® 5.0 demo (GRAPHPAD SOFTWARE INC., 2007).

It should be emphasised that the decision criterion used in this test was the p-value. If the p-value was greater than a, the null hypothesis formulated for the test would not be rejected, which consisted of the assertion that there were no statistical differences between the average rainfall intensity values of the river basins obtained using the arithmetic mean, Thiessen polygons and isohyets techniques. On the other hand, if the p-value is lower than the established a, this would mean that the null hypothesis should be rejected in favour of the alternative hypothesis, as there are statistical differences between the results provided by the techniques used.

It should also be noted that the Kruskal-Wallis test was complemented by the non-parametric Dunn's test, also carried out using Graphpad Prism® 5.0 demo (GRAPHPAD SOFTWARE INC., 2007), in order to ascertain in which of the techniques the discrepancies would lie, if any, which would force the rejection of the null hypothesis.

CHAPTER 3

RESULTS AND DISCUSSION

3.1 IDF CURVES

In order to check the applicability zones of the IDF equations, Figure 6 was constructed for the state of Rondônia, whose constants for each location are shown in Table 5. Analysing this figure, it can be seen that practically the entire state was covered, with the exception of a small area in the far north, belonging to the municipality of Porto Velho. This region was not included in the areas of influence of the IDF curves due to the spatial layout of the stations used, which made it impossible to include the others in the interpolation phase carried out by the software.

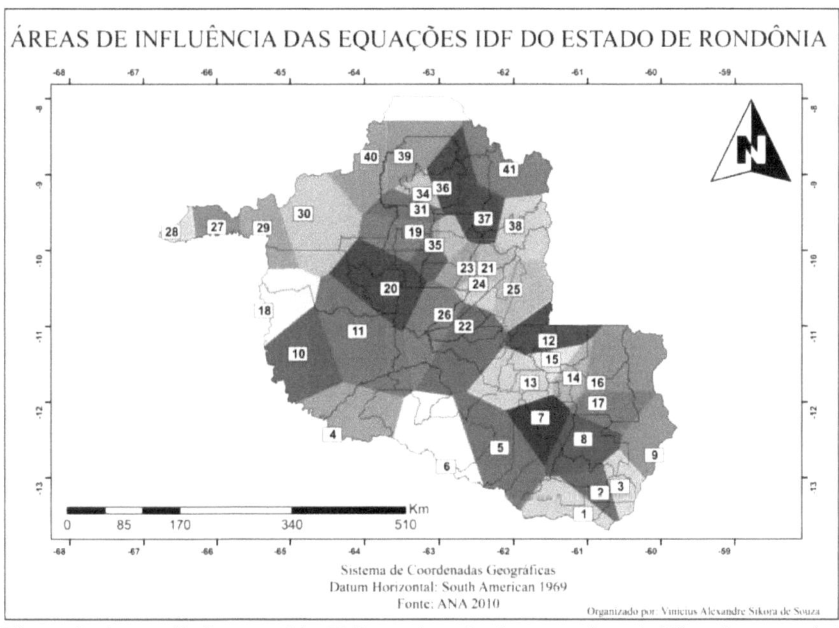

Figure 6 - Areas of influence of the IDF equations for the respective rainfall stations sampled in the state of Rondônia.

For this reason, it is recommended that the equation for area of influence 39 be adopted for the northern part of the municipality of Porto Velho, since this is the closest geographically to the location with no data and, possibly, both regions have similar climatic characteristics, producing a smaller error when using this equation.

The constants of the general form of the IDF equation (Equation 7) presented in Table 5 were calibrated in this study using the intensity of maximum rainfall with durations of 5, 10, 15, 20, 30, 60, 120, 480, 600, 720 and 1,440 min; and return periods of 2, 5, 10, 20 and 100 years. It therefore has a wide range of application due to the amplitude of the intervals, so this estimate can be used in various hydraulic works.

In general, it is possible to observe that the proposed equations show the typical behaviour for IDF curves, i.e. intensity is indirectly proportional to duration, as Pereira, Silveira and Silvino (2007) note when they observe that the shorter the duration of precipitation, the greater the average intensity.

A directly proportional relationship was found between the intensity and the return period, thus highlighting the difference in the use of Tr for sizing hydraulic works, since high return periods indicate high rainfall intensity. In this way, the design of more complex hydraulic projects requires the forecasting of hydrological quantities of great magnitude with a long recurrence interval, in order to estimate the maximum flows or precipitation that may occur in a certain location. In this context, the cost of such a project is inextricably linked to the period of recurrence of the phenomenon (BEIJO; MUNIZ; CASTRO NETO, 2005).

Table 5 - Coefficients of the IDF equations for the state of Rondônia.

Equation	Code	K	a	b	c
1	1360002	865,211	0,284	13,868	0,714
2	1360001	922,522	0,242	13,868	0,714
3	1360000	926,887	0,213	13,868	0,714
4	1264000	786,164	0,244	13,868	0,714
5	1262001	822,730	0,220	13,868	0,714
6	1262000	829,215	0,206	13,868	0,714
7	1261001	801,558	0,205	13,868	0,714
8	1261000	763,368	0,247	13,868	0,714
9	1260001	734,652	0,233	13,868	0,714
10	1164001	684,101	0,271	13,868	0,714
11	1164000	731,896	0,146	13,868	0,714
12	1161003	778,344	0,172	13,868	0,714
13	1161002	895,658	0,204	13,868	0,714
14	1161001	867,613	0,183	13,868	0,714
15	1161000	941,599	0,231	13,868	0,714
16	1160002	819,248	0,203	13,868	0,714
17	1160000	940,299	0,226	13,868	0,714
18	1065002	935,889	0,239	13,868	0,714
19	1063001	973,900	0,207	13,868	0,714
20	1063000	1038,794	0,293	13,868	0,714
21	1062004	873,381	0,196	13,868	0,714
22	1062003	742,118	0,189	13,868	0,714
23	1062002	926,701	0,173	13,868	0,714
24	1062001	918,014	0,204	13,868	0,714
25	1061003	963,613	0,161	13,868	0,714

26	1061001	951,542	0,198	13,868	0,714
27	966001	699,870	0,208	13,868	0,714
28	966000	918,097	0,190	13,868	0,714
29	965001	930,547	0,314	13,868	0,714
30	964001	703,547	0,322	13,868	0,714
31	963009	875,778	0,244	13,868	0,714
32	963006	933,698	0,255	13,868	0,714
33	963004	945,314	0,188	13,868	0,714
34	963001	981,296	0,182	13,868	0,714
35	963000	1053,185	0,206	13,868	0,714
36	962001	1093,628	0,216	13,868	0,714
37	962000	945,329	0,174	13,868	0,714
38	961003	903,301	0,206	13,868	0,714
39	863003	851,672	0,177	13,868	0,714
40	863000	1774,359	0,332	13,867	0,714
41	862000	826,299	0,191	13,868	0,714

When analysing the regional coefficients obtained for Equation 7 (TABLE 5), it can be seen that the "b" constants for each location, with the exception of the 40th, are numerically the same. The same aspect was observed for the "c" constant, but this characteristic was found for all 41 regions that had their IDF curves equated. This numerical equality is due to the disaggregation method used to transform daily rainfall into rainfall events with shorter durations, which uses identical disaggregation quotients for all historical series.

This means that these regional parameters are functions of rainfall intensity relationships of different durations. Other studies have also described the same regional coefficients, such as the study by Oliveira et al. (2000). These authors noted a similar fact to the ones described above, and in their studies they used a methodology similar to the one presented in this work to estimate intense rainfall equations for some locations in the state of Goiás.

Table 6 shows the analysis of the residuals generated by each IDF equation when estimating the data measured and modelled by the Gumbel distribution, which were used in the process of making this mathematical model of extreme rainfall. It can be seen that the fluctuations in the results of the equations are not significant when compared to the data used in their estimation, as the average standard error rate (SEM) ranged from 3.215 to 32.382, in agreement with the values obtained for the average normalised errors (ANE), which are in the range of -13.359 to -5.816%.

The negative values found in the ENM show that the equations proposed in this study tend to underestimate the measured data, albeit to an insignificant degree. It should be noted that the work by Oliveira et al. (2000) showed average normalised errors of approximately 14.4% in relation to the IDF equations for some locations in Goiás developed in their study.

It should also be noted that the mean multiplicative error (MMA) indices, which according to Moog and Jirka (1998) are very sensitive to small disparities, were close to 1, which shows that there were no significant differences in the results provided by the mathematical models when compared to the measured data.

Table 6 - Results of the validation analyses of the IDF equations.

Equation	EPM	ENM (%)	EMM	r	r^2	99.5% confidence interval	
						Bottom	Superior
1	7,622	-7,341	1,124	0,995	0,990	-4,020	2,960
2	7,927	-7,286	1,123	0,995	0,990	-3,420	2,090
3	6,326	-6,598	1,113	0,996	0,992	-2,900	1,500
4	5,720	-6,427	1,111	0,996	0,993	-2,940	1,830
5	5,829	-6,436	1,111	0,996	0,993	-2,680	1,420
6	4,614	-6,507	1,112	0,996	0,993	-2,420	1,300
7	5,351	-6,452	1,111	0,996	0,993	-2,370	1,220
8	6,829	-7,435	1,125	0,995	0,990	-2,900	1,830
9	7,383	-6,979	1,119	0,996	0,991	-2,600	1,430
10	7,469	-8,406	1,137	0,994	0,988	-2,950	1,990
11	8,903	-7,723	1,129	0,995	0,989	-1,560	0,560
12	4,235	-6,229	1,107	0,997	0,994	-1,890	0,770
13	4,617	-5,954	1,102	0,997	0,994	-2,600	1,370
14	10,494	-9,073	1,145	0,993	0,986	-2,230	1,090
15	7,839	-7,201	1,122	0,995	0,991	-3,290	1,800
16	5,199	-6,429	1,111	0,996	0,993	-2,400	1,240
17	7,651	-6,655	1,114	0,996	0,992	-3,190	1,690
18	5,942	-6,736	1,115	0,996	0,992	-3,430	2,000
19	3,975	-6,041	1,103	0,997	0,994	-2,880	1,520
20	13,631	-9,656	1,152	0,993	0,985	-5,120	3,980
21	5,866	-7,042	1,120	0,996	0,991	-2,400	1,290
22	5,299	-6,241	1,107	0,997	0,993	-1,970	1,020
23	3,215	-5,861	1,100	0,997	0,995	-2,270	0,920
24	5,824	-6,363	1,109	0,997	0,993	-2,700	1,390
25	5,352	-6,137	1,105	0,997	0,994	-2,210	0,850
26	7,098	-6,855	1,117	0,996	0,992	-2,670	1,420
27	4,774	-6,145	1,106	0,997	0,994	-2,090	1,100
28	11,821	-12,215	1,179	0,991	0,981	-2,460	1,280
29	14,671	-11,420	1,171	0,991	0,982	-5,370	4,490
30	5,138	-6,460	1,111	0,996	0,993	-4,370	3,650
31	32,282	-13,359	1,190	0,990	0,980	-3,290	2,030
32	4,811	-6,247	1,108	0,997	0,993	-3,720	2,440
33	5,359	-6,206	1,107	0,997	0,994	-2,490	1,290
34	4,765	-6,051	1,104	0,997	0,994	-2,490	1,190
35	6,822	-6,483	1,111	0,996	0,993	-3,110	1,620
36	4,906	-6,077	1,104	0,997	0,994	-3,460	1,820
37	6,333	-6,486	1,111	0,996	0,993	-2,330	0,950
38	5,854	-6,467	1,111	0,996	0,993	-2,650	1,410
39	4,491	-6,088	1,104	0,997	0,994	-2,110	0,940
40	6,838	-7,333	1,124	0,995	0,990	-11,600	9,700
41	5,242	-6,304	1,109	0,997	0,993	-2,220	1,170

It is worth pointing out that the equations proposed in this study showed a regression coefficient (TABLE 6) of approximately 0.980 to 0.995, indicating that 98 to 100 per cent of the variations in intensity data are explained by variations in the duration and return period. Thus, the correlation coefficient (r) of these estimates is in the range of approximately 1, demonstrating, according to Levin (1977) cited by Elias et al. (2009), that the relationship of i_m is perfectly positively correlated with the other two variables.

The Wilcoxon-Mann-Whitney hypothesis test showed that there was no statistical evidence that the data estimated by the equation differed from the measured data, since the lower and upper limits of the confidence intervals had a value of 0, so the null hypothesis was not rejected. It can therefore be seen that the extrapolation of data to a period of 100 years in the majority of the historical series did not cause distortions of any great magnitude that could jeopardise the estimation of these equations, even though some of the available data had a maximum return period of less than 10 years. Thus, it can be affirmed with 99.5% confidence that the IDF equations proposed in this study are significant for the data used, which confirms the feasibility of using them for the processes they are intended for.

3.2 PROBABILITY DISTRIBUTION

The results of the adherence assessments are summarised in Table 7, using the Kolmogorov-Smimov test to check the fit of the station data to the empirical probability distributions. As explained above, the Weibull++ 7® software (RELIASOFT, 2011) was used.

This programme presents the results in the following format: it makes a relationship between the parameters D_{er} it and D_{ca} ic, and relationships that are less than 1 indicate that the theoretical distribution tested can be used to predict the behaviour of the data observed by a given rainfall station, since values of this magnitude express the statement that the parameter Dcrit is greater than the parameter D_{ca} ic, which is the criterion for accepting the null hypothesis (Ho).

However, if this relationship has a value of 1, this indicates that the statistical distribution being analysed does not correspond to the probabilistic distribution of the maximum rainfall data, since the statement D_{cr} it < D_{ca} ic is true, which means that the null hypothesis must be rejected and the alternative hypothesis accepted.

Analysing Table 7, it can be seen that among the distributions analysed, only the Gamma distribution did not correspond to the distribution of the data in all the cases tested by KS.

Thus, the results obtained differ from Naghettini and Pinto (2007), who state that the versatility of the shapes, the variable and positive asymmetry coefficient, together with the fact that the random variable is not defined for negative values, make the Gamma distribution a very attractive probabilistic model for representing hydrological and hydrometeorological variables. Furthermore, this finding partially rules out the efficiency of the Gamma distribution described by Haan (1977) in modelling daily, weekly, monthly and annual precipitation heights.

Table 7 - Results of the adherence tests using KS and symmetry analysis.

| Station | Dcalc/Dcrit | | | | | | | n | Coefficient of Symmetry |
	Normal	Log-Normal	Gama	Weibull	Logistics	Exponential	Gumbel		
1	0,706	0,997	1,000	0,971	0,647	1,000	0,900	28	0,451
2	0,348	0,028	1,000	0,213	0,265	1,000	0,736	28	0,608
3	0,230	0,284	1,000	0,186	0,130	1,000	0,573	28	0,477
4	0,254	0,210	1,000	0,209	0,397	1,000	0,676	28	0,192
5	0,927	0,500	1,000	0,858	0,897	1,000	0,987	12	-0,242
6	0,483	0,729	1,000	0,276	0,574	1,000	0,116	31	-0,568
7	0,008	0,053	1,000	0,004	0,022	0,999	0,019	12	-0,041
8	0,805	0,696	1,000	0,872	0,863	1,000	0,923	21	0,642
9	0,381	0,010	1,000	0,108	0,292	1,000	0,616	24	2,802
10	0,010	0,551	1,000	0,158	0,002	0,998	0,128	19	0,215
11	0,000	0,000	1,000	0,000	0,000	1,000	0,001	12	0,018
12	0,036	0,024	1,000	0,072	0,017	1,000	0,096	12	-0,126
13	0,045	0,012	1,000	0,107	0,082	1,000	0,530	28	0,117
14	0,296	0,081	1,000	0,557	0,457	1,000	0,738	31	0,575
15	0,177	0,203	1,000	0,119	0,136	1,000	0,706	34	0,999
16	0,384	0,089	1,000	0,611	0,482	1,000	0,826	29	2,050
17	0,340	0,004	1,000	0,348	0,444	1,000	0,822	34	0,511
18	0,927	0,500	1,000	0,858	0,897	1,000	0,987	39	2,015
19	0,295	0,026	1,000	0,458	0,418	1,000	0,766	29	0,577
20	0,996	0,834	1,000	0,953	0,996	1,000	0,999	31	3,714
21	0,418	0,262	1,000	0,550	0,552	1,000	0,669	24	0,192
22	0,061	0,449	1,000	0,020	0,028	1,000	0,254	28	-0,288
23	0,024	0,288	1,000	0,005	0,006	1,000	0,206	33	0,292
24	0,269	0,009	1,000	0,192	0,206	1,000	0,731	34	1,892
25	0,179	0,099	1,000	0,385	0,151	1,000	0,635	23	-0,001
26	0,045	0,000	1,000	0,046	0,018	1,000	0,350	23	0,760
27	0,012	0,242	1,000	0,000	0,026	1,000	0,022	28	-0,459
28	0,879	0,640	1,000	0,847	0,840	1,000	0,963	34	1,518
29	0,903	0,262	1,000	0,139	0,869	0,998	0,959	33	4,368
30	0,415	0,969	1,000	0,879	0,424	0,996	0,176	24	-0,117
31	0,021	0,330	1,000	0,155	0,007	1,000	0,002	14	-0,651
32	0,299	0,708	1,000	0,388	0,335	1,000	0,693	19	-0,555
33	0,471	0,097	1,000	0,650	0,539	1,000	0,875	31	0,793
34	0,504	0,140	1,000	0,455	0,389	1,000	0,786	34	1,113
35	0,292	0,002	1,000	0,282	0,231	1,000	0,785	36	0,829
36	0,206	0,091	1,000	0,201	0,187	1,000	0,695	27	1,043
37	0,007	0,144	1,000	0,106	0,048	1,000	0,311	29	0,598
38	0,492	0,192	1,000	0,473	0,889	1,000	0,770	21	0,821
39	0,466	0,181	1,000	0,605	0,481	1,000	0,806	19	0,866
40	0,999	0,959	1,000	0,998	1,000	0,999	0,990	45	1,752
41	0,019	0,361	1,000	0,006	0,007	1,000	0,242	34	-0,070

The analyses carried out, shown in Table 7, were of one-day annual maximum time series. Thus, the results presented above on the gamma distribution show that it is only suitable for modelling short periods of rainfall, as concluded by Thom (1958, apud MURTA et al., 2005), and as evidenced in the studies by Moreira et al. (2010); Murta et al. (2005); Silva et al. (2007); and Vasconcellos,

André and Perecin (1999).

The Exponential distribution as well as the Gamma distribution showed a high rejection rate in terms of their fit with the sample data used, and their use was only recommended for 5 of the 41 stations evaluated. This rejection phenomenon was also observed by Catalunha et al. (2002) when they tested this distribution with rainfall data from the state of Minas Gerais.

The authors attributed the modelling of this hydrological element to the inefficiency of the Exponential distribution. The authors also emphasised the overestimation that this statistical model causes in most rainfall data. In this way, the KS showed that most of the data does not have an asymmetry in the shape of an inverted "J", a typical characteristic of the Exponential distribution that would approve its use (CATALUNHA et al., 2002).

Table 7 shows that there was no rejection of the adherence of the data to the Weibull, Log-Normal, Gumbel and Logistic distributions. In this way, the result of the Weibull distribution agreed with the statement by Catalunha et al. (2002), who highlighted the use of this statistical model in hydrological analysis for extreme events, as well as its superiority when compared to the Gamma distribution for modelling rainfall data, as can be seen above.

For the static Log-Normal and Gumbel models, the work of Back (2001) confirms the above findings, as the author, when selecting a statistical distribution to describe extreme rainfall in the state of Santa Catarina among the various models, concluded that the Gumbel and Log-Normal distributions presented the best fit for most of the rainfall stations studied.

With regard to acceptance of the Log-Normal distribution, in all cases, unlike the Gamma distribution, it was revealed that the probabilities of the rainfall data analysed are arranged in the cores of the distribution curves and not in their falls, as Tucci (2009) points out that the falls of the Log-Normal and Gamma distributions are similar, which would lead to acceptance of the same nature, given the asymmetrical characteristics of these statistical models.

The results obtained by the Logistic distribution showed that it can be used to model extreme rainfall events for the data used from the state of Rondônia, and the KS showed that this distribution can be applied to other areas, in addition to those mentioned in the work by Bittencourt (2001), such as health, economics, administration and education.

An analysis of Table 7 shows that all the data was accepted for the Normal distribution, even though the data was asymmetrical, a characteristic that would automatically refute the null hypothesis for this empirical distribution, since the symmetrical arrangement of the data is a latent attribute of such a distribution (NAGHETTINI; PINTO, 2007).

This abnormality can be explained by the inferiority of the KS test in confirming the normality of the data. This finding was made in the work by Leotti, Birck and Ribold (2005), where, when assessing data that did not contain normality, the Kolmogorov-Smimov test gave untruthful answers, stating that the data was normal, and the KS, of all the tests analysed, was the one with the worst performance when it came to assessing normal distribution.

The results in Table 7 generally show that the answers provided by the KS test should be analysed rigorously, as it was highly acceptable for most distributions. This observation is similar to the findings of Catalunha et al. (2002), who found that the Kolmogorov-Smimov test's approval level for different probability distributions is high, thus generating insecurity in its use.

The aforementioned author attributed this phenomenon of high acceptance to the asymmetric conditions of statistical distributions, since this characteristic gives higher values in the initial classes and lower values in the final ones, which jeopardises the KS test because it measures the differences between theoretical and empirical probabilities.

In this sense, Araújo et al. (2007) also verified this high acceptance range of KS when they analysed the adherence of rainfall data from some stations in the Lagoa Mirim basin in Rio Grande do Sul, in relation to the distributions: Normal, Log-Normal, Log-Pearson III and Gumbel.

However, these same authors pointed out that when they plotted the data on appropriate probability paper, the observed data did not adhere adequately to the theoretical model, emphasising once again that the results provided by KS should be used sparingly.

However, in view of the above, it should be emphasised that according to Naghettini and Pinto (2007) the adherence test is only one of three tools that should be taken into account when selecting an empirical probabilistic distribution to represent hydrological data, the other two being: the physical characteristics of the phenomenon in question and the possible theoretical deductions regarding the distributive properties of the variable in question.

In these terms, the literature shows that the Gumbel distribution is the most suitable for representing extreme rainfall events, as Hershfield and Kohler (1960) analysed data from thousands of rainfall stations in the United States and proved that the Gumbel distribution is more efficient at describing the phenomenon of heavy rainfall.

In this context, the works by Beijo et al. (2003); Oliveira et al. (2000); Oliveira et al. (2008); and Silva et al. (2003), are references in the use of the Gumbel distribution for modelling extreme rainfall, thus proving the applicability of this distribution to this phenomenon, which is why the Gumbel distribution was chosen to estimate the IDF curves for this study.

3.3 EXTREME RAINFALL

The extreme rainfall events recorded in the historical series of the stations sampled in this study are shown in Table 8, which shows the intensities (i) of the maximum and minimum annual events, with their respective years of occurrence (AO) and return periods (Tr). The table also contains the averages of the maximum events and the standard deviation for this phenomenon.

In general, the average maximum "one-day" rainfall in the state of Rondônia ranges from approximately 71.3 to 173.5 mm/day, both averages being found at stations located in the municipality of Porto Velho.

This high rate of rainfall observed in all the historical series of the stations sampled is characteristic of the climatic aspects of the region, since according to Rondônia (2009), the average annual rainfall in the state of Rondônia varies between 1,400 and 2,600 mm/year.

Table 8 - Extreme events at the sampled stations.

| Station | Maximum | | | Minimum | | | Average (mm/day) | Standard Deviation (mm/day) |
	i (mm/day)	Tr (years)	AO (year)	i (mm/day)	Tr (years)	AO (year)		
1	227,4	55,906	2008	9,7	1,009	2002	90,860	43,338
2	176,6	19,551	1996	42	1,039	1986	98,025	35,978
3	165	23,387	1992	52,5	1,028	1989	98,728	28,199
4	135	12,037	1998	28,9	1,010	2004	83,885	29,553
5	123,4	9,455	2010	48,5	1,034	2000	86,391	21,562
6	126,4	13,158	1998	39,5	1,000	1990	89,019	20,849
7	110,5	6,975	2007	57,4	1,101	2000	84,041	19,087
8	144	16,918	1985	48	1,177	1994	80,559	30,037
9	195,5	128,250	1981	41,4	1,135	1990	76,712	29,869
10	144,8	23,254	1992	10	1,005	1998	71,578	29,726
11	95	10,966	1986	56	1,023	1985	75,416	10,443
12	109	12,339	2006	52,4	1,010	1999	81,058	13,965
13	141	13,606	1991 1992	43,6	1,002	2004	95,432	24,695
14	131,2	13,919	2003	62,4	1,071	1986	91,738	21,385
15	215,3	66,309	1990	21,2	1,000	1989	100,161	35,531
16	184,2	110,170	1997	52,2	1,090	1982	86,251	26,088
17	173,4	21,306	1998	35,4	1,004	1989	100,438	32,006
18	222	52,721	1987 2003	41,2	1,061	1990	98,607	41,080
19	167,4	20,378	1996	58	1,035	1995	103,537	28,848
20	423,4	285,350	2002	31,8	1,184	1992	104,068	69,667
21	134,2	13,676	1998	59,5	1,051	2008	92,650	22,037
22	115,4	18,010	1987	32,7	1,000	1994	78,817	17,357
23	150,6	33,308	1979	60	1,007	1997	97,924	20,033
24	212,8	147,970	1983	53,3	1,045	2010	97,008	29,259
25	140,8	19,844	2005	65	1,005	1989 2003	100,921	17,713
26	170	31,259	1991	58,7	1,035	1996	100,556	25,907
27	101,2	8,756	1988	30,1	1,000	2004	74,939	18,414
28	190	97,333	1993	46	1,005	1990	96,867	25,988
29	464,8	536,010	2007	28	1,280	1990	90,387	73,169
30	151	13,805	1993	9,4	1,057	1997	71,341	42,048

31	135,7	8,764	2003	27	1,001	1997	92,128	27,478
32	147,7	8,271	1996	14,3	1,000	2000	98,794	33,839
33	165,2	33,474	1996	59,8	1,027	1980	99,977	24,600
34	185	67,780	1984	64	1,032	2002	103,567	24,978
35	201,2	39,063	1990	63,5	1,043	2008	112,114	32,391
36	218	38,680	1991	53,2	1,019	1986	115,855	36,152
37	150	24,404	1978	69,3	1,060	2007	99,606	21,172
38	150,2	16,101	2005	47	1,015	1992	95,300	26,429
39	133,2	19,669	1992	60,4	1,058	1993	89,147	19,146
40	665,3	77,784	2001	43,2	1,257	1989	173,542	149,014
41	137,8	27,204	1981	40,8	1,000	1991	88,088	20,381

Table 8 shows that the highest level of intense rainfall found in all the historical series was 665.3 mm/day in Porto Velho in 2001, but the period for a phenomenon of the same magnitude or greater to occur again in this location is only 77.784 years. This shows that the area in which this rain gauge is located has a high rate of extreme rainfall, a fact that can be observed by the return period.

However, Carvalho (2007) found a maximum height of 128.1 mm/day of Tr of 18.99 years in 1999, when he analysed a 34-year series of intense rainfall in the Igarapé Murupu catchment area in Boa Vista - RR, which belongs to the same region as the area in this study.

The event with the longest period of recurrence shown in Table 8 was in the municipality of Porto Velho, with 536.010 years and a magnitude of 464.8 mm/day, which was observed in 2007.

It should be noted that the occurrence of this phenomenon was even picked up by the rainfall estimation radars of the Centre for Weather Forecasting and Climate Studies of the National Institute for Space Research (CPTEC/INPE), given the core of the storm, which moved northwest/southeast, with the north and west of Porto Velho being the most affected by the storm (DE OLHO NO TEMPO, 2011).

Table 8 shows the occurrence of maximum intense rainfall in some rainfall stations during the same period, with the years 1992, 1991, 1998, 1996 and 2003 having the most repetitions of maximum extreme intensities.

Therefore, during these periods it is clear that some meteorological phenomenon is at work, such as *La Nina*, which according to Cutrim, Molion and Nechet (2000) causes positive rainfall anomalies in the Amazon region. However, this same meteorological phenomenon was pointed out by Bergamaschi et al. (2004) as being responsible for the occurrence of rainfall below the historical average in southern Brazil.

With regard to the minimum annual extreme events, these show the existence of drier periods in certain regions, as it can be seen that the occurrences are well below the averages for the same locations, which may indicate the presence of some climatic phenomenon that interfered significantly with the rainfall in those years, such as *El Nino*, highlighted in the study by Silva, Saraiva and Silva (2010) as the main factor in the droughts recorded in the Amazon region in the periods 1925-1926, 1968-1969 and 1997-1998.

In this sense, Santos and Buchmann (2010) report that this phenomenon is characterised by the warming of the surface waters of the equatorial Pacific Ocean (central-western portion) and the weakening of the easterly trade winds. As such, it can cause climate change and financial losses in various parts of the world.

Therefore, the findings explained above are in line with the results of Grimm and Tedeschi (2004), who analysed daily rainfall data from 2714 weather stations across Brazil, sourced from the ANA, during the period 1956-2002, and found that in the northern region of the country, there is a greater increase in extreme events during *La Nina* and a decrease during *El Nino, although* their behaviour is not very symmetrical, as *La Nina* causes increases in rainfall more often than *El Nino causes a* reduction in this variable.

It should be emphasised that the discrepancies in the extreme maximum and minimum heights justify the high values found for the standard deviations of the series. This high dispersion of rainfall data, indicated by the standard deviation, was also shown by Mehl et al. (2001), who, when characterising rainfall patterns in Santa Maria - RS, observed standard deviations of over 30 mm in rainfall lasting 10 min.

Another notable factor is the lack of a pattern between the extreme maximum and minimum intensities for the various locations where the analysed rain gauges are located. Two elements are responsible for this characteristic: the historical series covering different periods, as shown in Table 2; and the meteorological phenomena present in the state that act on the dynamics of the rainfall regime, such as high daytime convections, the Bolivian High, the Intertropical Convergence Zone and the Lines of Instability (GAMA, 2003).

Figure 7 shows the distribution of the most intense extreme rains with the greatest recurrence in Rondônia, with a duration of 5 minutes and a return time of 2 years, in the form of isohyets. It should be noted that rainfall of these characteristics is pointed out by Brandão, Rodrigues and Costa (2001) as rainfall with a high erosive factor, which can cause environmental and economic damage. This illustration also shows the characteristics of land use and occupation in the state, according to data collected by RADAM (BRASIL, 1978).

It is possible to see in Figure 7 that the extreme rainfall with the attributes described above increases in intensity as it approaches the northernmost region of the state. This may be due to the greater proximity to the equator, where a low-pressure zone called the Intertropical Convergence Zone (ITCZ) is present.

According to Xavier et al. (2000), this phenomenon favours the formation of cumulus clouds and is responsible for heavy rainfall in tropical regions such as Brazil. In this context, the work by Xavier et al. (2003) emphasised that the higher intensities recorded in the Ceará river basins are due to the action of the ITCZ.

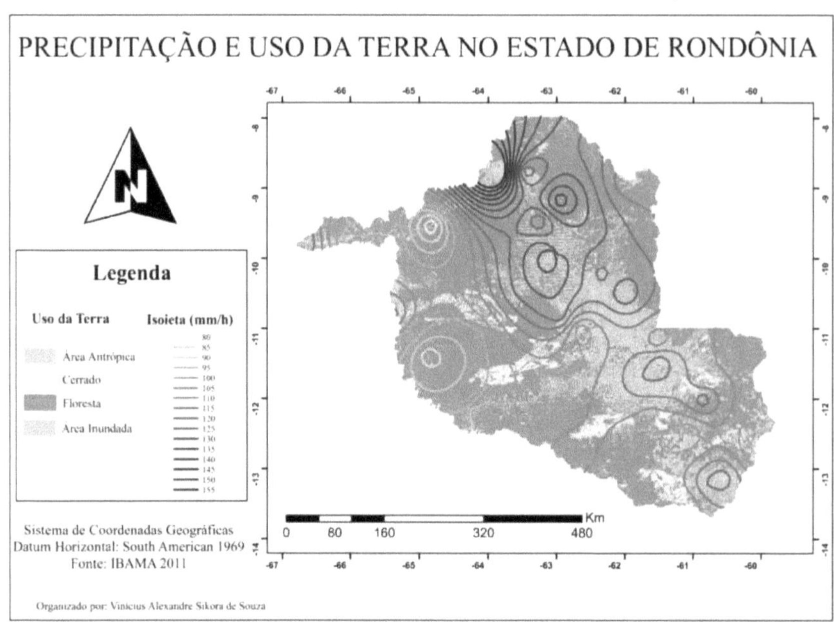

Figura 7 - The relationship between maximum rainfall and land use and occupation in the state of Rondônia.

When relating land use and occupation to rainfall, there is no clear relationship between predominant land use and the minimisation or maximisation of extreme rainfall events.

The explanation for this may lie in the lack of understanding of the hydrological cycle of the Amazon region in terms of the influence of deforestation, as Cohen et al. (2007) explain the lack of consensus, in the use of General Circulation Models (GCMs), that deforestation in the Amazon region causes a generalised decrease in precipitation in this same region.

However, when examining the southeastern region of Rondônia (FIGURE 7), there is a slight reduction in rainfall intensity, which may be due to the presence of several urban centres in this region.

This is in line with the study by Werth and Avissar (2002), in which they carried out six climate simulations and observed sharp reductions in precipitation, evapotranspiration and cloudiness as a result of deforestation in the Amazon. According to the aforementioned authors, these changes are detectable all over the planet.

Thus, it can be observed that the transformation of forested areas into urbanised centres can influence the dynamics of rainfall distribution. (2004, apud COHEN et al., 2007), alterations to the water, solar energy, carbon and nutrient cycles resulting from changes in land use in the Amazon region can have climatic and environmental consequences on a local, regional and global scale.

In this context, the work by Santos Neto, Moraes and Nóbrega (2010) addresses a notable decline in the average decadal rainfall values in this same area, analysing the impact of anthropogenic action in the central region of Rondônia, using a historical rainfall series from 1951 to 2008.

The influence of deforestation on rainfall in the state of Rondônia has also been observed by Butt, Oliveira and Costa (2011), who emphasise the existence of a marked change in the start time of the state's rainy season, projecting delays of 18 days over the next 30 years if the characteristics of land use change continue.

Table 9 shows the average maximum rainfall intensities for the seven river basins in the state of Rondônia. These events have the same characteristics as in Figure 7, with a duration of 5 minutes and a return time of 2 years.

Table 9 - Average maximum intensities for the Rondônia Hydrographic Basins, calculated using the following methodologies: Arithmetic Mean (A.M.), Isoietas and Thiessen Polygons (T.P.).

River Basin	Average Intensity (mm/h)		
	M. A.	Isoietas	P. T.
Guaporé River Basin	107,329	103,728	107,176
Mamoré River Basin	98,103	95,540	95,732
Abunã River Basin	104,170	101,476	113,624
Madeira River Basin	118,956	109,820	110,874
Jamari River Basin	120,390	124,634	110,864
Machado River Basin	114,567	113,121	114,363
Roosevelt River Basin	96,758	103,667	117,929

By relating the results of the average extreme rainfall of the Rondônia Hydrographic Basins, obtained by the Arithmetic Mean (A.M.), Thiessen Polygons (T.P.) and Isoietas methods, it can be seen that for most of the basins analysed the distribution values of this variable were close in the different methods.

According to Braz et al. (2007), this finding indicates that the spatial layout of the rain gauge stations analysed is homogeneous in most river basins (BRAZ et al., 2007). As an exception to this fact, the Roosevelt River Basin showed the most discrepant values, due to the low density of rain gauges in this area (TIBÚRCIO; CASTRO, 2007). In addition, this fact may also have occurred due to the fact that this same region has more abrupt elevation transitions, as can be seen in Figure 8, thus generating a more rugged terrain which, according to Tucci (2009), makes the average intensity values of the watershed obtained by the P.T. technique differ from those measured by the isohyets.

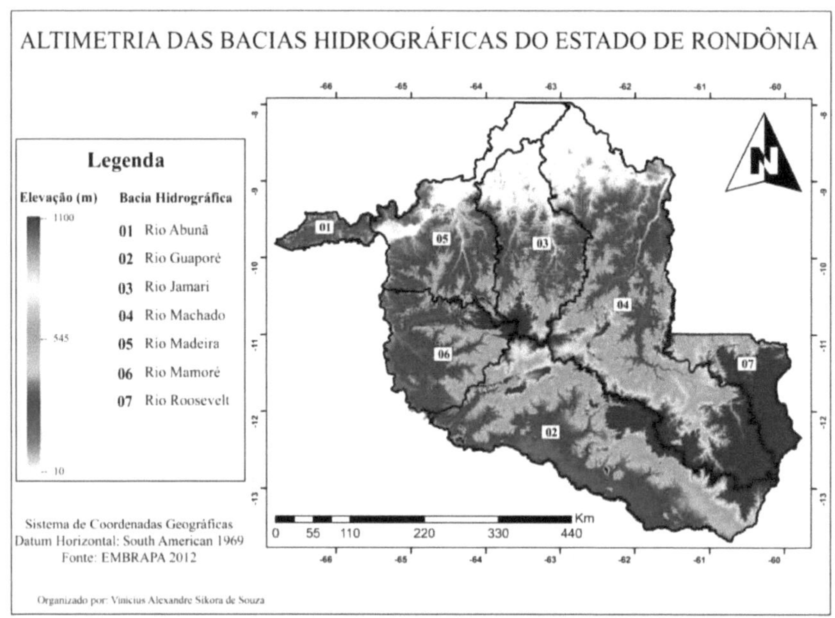

Figura 8 - Elevation of the hydrographic basins in the state of Rondônia.

In this respect, the regions of the Machado and Madeira river basins, due to their extensive plains, showed the closest average intensity values using these techniques.

However, it should be noted that the anomalies observed in Table 9 are not statistically significant, as the p-value obtained for the K-W test was 0.660, which is much higher than the established significance level, which means that the results obtained by the different techniques are statistically identical, a similar result was shown by the Dunn test, as it indicated no differences in its pair analyses, thus reinforcing that the differences observed are not statistically significant.

Comparing Figure 7 with Figure 8 shows that the distribution of maximum rainfall in the state of Rondônia shifts towards the lower altitudes. This finding disagrees with the trends observed in the study by Costa et al. (2008), where the authors analysed this phenomenon in Portugal and found that the distribution of maximum rainfall varied locally with elevation, with the highest elevation areas receiving the most rainfall. It can thus be seen that unlike Portugal, where extreme rainfall is caused by orographic rains, in the state of Rondônia this is a function of the ITCZ, as mentioned above.

The data shown in Table 9 also reveals that the Jamari River Basin has extreme events of greater average magnitude when compared to the other river basins, thus indicating that from the point of view of the total volume of rainfall, there is a need for greater concern with flood control and soil conservation in this part of the state of Rondônia (CARDOSO; ULLMANN; BERTOL, 1998).

Therefore, given the characteristics of the rainfall analysed, this watershed is more likely to suffer changes in the characteristics of the pastures; the loss of quality of water bodies, due to the loading of sediments that harm the life of aquatic fauna species; the deterioration of dirt roads; and loss of habitats and loss of local biodiversity (CARVALHO, 2008).

CHAPTER 4

FINAL CONSIDERATIONS

This work has produced many relevant results, since it has constructed intense rainfall equations that can be used to support the design of hydraulic works and hydrological studies.

The study confirms that the equations proposed for estimating the intensity of maximum rainfall are highly applicable to the regions of the municipalities of Rondônia, showing a high degree of correlation with the related variables, duration and period of return of the phenomenon. This finding was made on the basis of statistical analyses and comparisons with data from the literature, and it was also possible to verify the high degree of reliability of the IDF curves.

The work showed that the KS adherence test generally provided significant acceptance for most of the statistical distributions tested for rainfall data. However, when relating the physical characteristics of extreme rainfall phenomena and the theoretical deductions regarding their distributional properties, it was seen that the Gumbel distribution is more suitable for modelling these events.

The historical series used revealed that the average maximum "one-day" rainfall in the state of Rondônia ranged from approximately 71.3 to 173.5 mm/day. In addition, it was observed that possible climatic phenomena such as *El Nino* and *La Nina* in the region analysed may have generated reductions and increases in the occurrence of extreme rainfall events, respectively.

This study found that rainfall events with a high erosive factor, a duration of 5 min and a return time of 2 years, increase in intensity as they approach the northernmost region of the state, indicating that they are influenced by the Intertropical Convergence Zone. Also in relation to these events, it was found that the Jamari River Basin, when compared to the other river basins in the state of Rondônia, has extreme rainfall of greater average magnitude, thus indicating the need for greater concern with flood control and soil conservation in this part of the state.

For future work, it is recommended that rainfall stations with records shorter than one day be installed in order to confirm the results obtained, since the rainfall data shorter than one day was obtained synthetically, so it may not correspond faithfully to the characteristics of the region under study.

In addition, it is recommended that rain gauge stations be installed in the northernmost region of Rondônia, because due to the lack of rain gauge stations in this region, it was not considered as an area of influence by any of the estimated equations.

REFERENCES

AGUIAR, R. G.; VON RANDOW, C.; PRIANTE FILHO, N.; MANZI, A. O.; AGUIAR, L. J. G.; CARDOSO, F. L. Mass and energy fluxes in a tropical forest in south-western Amazonia. **Revista Brasileira de Meteorologia,** v. 21, p. 248-257,2006.

ALMEIDA, A. Q.; RIBEIRO, A.; PAIVA, Y. G.; RASCON, N. J. L.; LIMA, E. P. Geostatistics in the study of temporal modelling of precipitation. **Brazilian Journal of Agricultural and Environmental Engineering,** v. 15, n. 4, p. 354-358, 2011.

ANIDO, N. M. R. **Hydrological characterisation of an experimental watershed with a view to identifying environmental monitoring indicators.** Piracicaba: USP, 2002. Dissertation (Master's Degree in Forest Resources), Escola Superior de Agricultura "Luiz de Queiroz", Universidade de São Paulo, 2002.

ARANTES, E. J.; PASSIG, F. H.; CARVALHO, K. Q.; KREUTZ, C.; ARANTES, E. A. Analysis of heavy rainfall in the north-western region of Paraná. **OLAM - Ciência & Tecnologia,** n. Esp., p. 31-47, 2009.

ARAÚJO, R. D.; DAMÉ, R. C. F.; TEIXEIRA, C. F. A.; ROSSKOFF, J. L.; TERRA, V. S. S.; GAYER, C. A. P.; FRAGA, D. S. Performance of the Kolmogorov-Smimov test in the selection of probabilistic theoretical models for predicting probable monthly rainfall in the Lagoa Mirim/RS basin. In: ENCONTRO DE PÓS-GRADUAÇÃO, 9., 2007, Pelotas. Proceedings..., Pelotas: FAEM, 2007.

ARGENTO, M. S. F.; CRUZ, C. B. M. Geomorphological Mapping. In: CUNHA,S. B.; GUERRA, A. J. T. (org.) **Geomorphology - Exercises, Techniques and Applications.** Ed. Bertrand Brasil SA, Rio de Janeiro, 1996, p. 239-249.

BACK, A. J. Selection of a probability distribution for extreme daily rainfall in the state of Santa Catarina. **Revista Brasileira de Meteorologia,** v. 16, n. 2, p. 211-222, 2001.

BACK, Álvaro J. Relationships between intense precipitation events of different durations in the municipality of Urussanga, **SC. Revista Brasileira de Engenharia Agrícola e Ambiental,** v. 13, n. 2, p. 170-175, 2009.

BARBOSA, S. E. S.; BARBOSA JÚNIOR, A. R.; SILVA, G. Q.; CAMPOS, E. N. B.; RODRIGUES, V. C. Generation of regionalisation models for maximum, long-period average and seven-day minimum flows for the Carmo river basin, Minas Gerais, Brazil.
General. **Engenharia Sanitária Ambiental,** Rio de Janeiro, v. 10, n. 1, Mar. 2005.
BEIJO, L. A.; MUNIZ, J. A.; CASTRO NETO, P. Return time of maximum precipitation in Lavras (MG) by the type I extreme value distribution. **Ciência Agrotécnica,** Lavras, v. 29, n. 3, p. 657-667, 2005.

BEIJO, L. A.; MUNIZ, J.A.; VOLPE, C. A.; PEREIRA, G. T. Study of maximum precipitation in Jaboticabal (SP) by the Gumbel distribution using two methods of parameter estimation. **Revista Brasileira de Agrometeorologia,** Santa Maria, v. 11, n. 1, p. 141-147, 2003.

BERGAMASCHI, H.; DALMAGO, G. A.; BERGONCI, J. L; BIANCHI, C. A. M.; MÚLLER, A. G.; COMIRAN, F.; HECKLE, B. M. M. Water distribution in the critical period of maize and grain production. **Pesquisa Agropecuária Brasileira, Brasília,** v. 39, n. 9, p. 831-839, 2004.

BITTENCOURT, H. R. **Comparison of logistic discrimination with the Gaussian maximum likelihood method in the classification of digital images.** Porto Alegre: UFGRS, 2001. Dissertation (Master's Degree in Remote Sensing), State Centre for Remote Sensing and Meteorology Research, Federal University of Rio Grande do Sul, 2001.

BOOTS, B. N. **Concepts and techniques in modern geography: Voronoi (Thiessen) polygons.** Norwich, UK: Geo Books, 1986.

BORGES, L. C.; FERREIR, D. F. Power and type i error rates of the Scott-Knott, Tukey and Student-Newman-Keuls tests under normal and non-normal residue distributions. **Revista de Matemática e Estatística,** São Paulo, v. 21, n. 1, p. 67-83, 2003.

BRANDÃO, C.; RODRIGUES, R.; COSTA, J. P. **Análise de fenómenos extremos precipitações intensas em Portugal Continental.** Lisbon: Directorate of Water Resources Services, 2001.

BRAZIL. National Department of Mineral Production. **RADAMBRASIL Project, Natural Resources Survey. Sheet SC - 20: Porto Velho.** Rio de Janeiro, v. 16, 1978.

BRAZ, R. L.; RIBEIRO, C. A. D.; FERREIRA, D. S.; CECILIO, R. A. Use of historical series and GIS techniques in the study of the temporal and spatial distribution of rainfall in the Barra Seca basin located in the north of the State of Espírito Santo. In: ENCONTRO LATINO AMERICANO DE INICIAÇÃO CIENTÍFICA, 11., 2007, São José dos Campos. Proceedings..., São José dos Campos: UNIVAP, 2007.
BUTT, N.; OLIVEIRA, P. A.; HEIL COSTA, M. H. Evidence that deforestation affects the onset of the rainy season in Rondônia, Brazil. **Journal of Geophysical Research,** v. 116, n. Dl 1120, p. 1-8, 2011.

CARDOSO, C. O.; ULLMANN, M. N.; BERTOL, I. Heavy rainfall analysis based on the disaggregation of daily rainfall in Lages and Campos Novos (SC). **Revista Brasileira de Ciência do Solo,** Viçosa, v. 22, n. 1, p. 131-140,1998.

CARVALHO, E. T. L. **Evaluation of rainwater infiltration elements in the northern zone of the city of Goiânia.** Goiânia: UFG, 2008. Dissertation (Master's Degree in Geotechnics and Civil Construction), School of Civil Engineering, Federal University of Goiás, 2008.

CARVALHO, R. G. **Obtaining the intensity-duration-frequency (IDF) curve of intense rainfall for the Igarapé Murupu Hydrographic Basin, Boa Vista - RR.** Boa Vista: UFRR, 2007. Dissertation (Master's Degree in Natural Resources), Postgraduate Programme in Natural Resources - PRONAT, Federal University of Roraima, 2007.

CASTRO, J. F. M. The importance of cartography in river basin studies. In: SEMANA DE ESTUDOS GEOGRÁFICOS "O HOMEM E AS ÁGUAS", 30, 2000, São Paulo. **Proceedings...** São Paulo: UNESP, 2000.

CATALUNHA, J. M.; SEDIYAMA, G. C.; RIBEIRO, A.; LEAL, B. G.; SOARES, C. P. B. Application of five probability density functions to rainfall series in the state of Minas Gerais. **Revista Brasileira de Agrometeorologia,** v. 10, n. 1, p. 153-162, 2002.

CECÍLIO, R. A.; PRUSKI, F. F. Interpolation of the parameters of the heavy rainfall equation using the inverse of powers of distance. **Revista Brasileira de Engenharia Agrícola e Ambiental,** v.7, n.3, p.501-504, 2003.

CECÍLIO, R. A.; XAVIER, A.C.; PRUSKI, F. F.; HOLLANDA, M. P.; PEZZOPANE, J. E. M. Evaluation of interpolators for the parameters of heavy rainfall equations in Espírito Santo. **Ambi-Agua,** Taubaté, v. 4, n. 3, p. 82-92, 2009.

CHOW, V. T.; MAIDMENT, D. R.; MAYS, L. W. **Applied hydrology.** Singapore: McGraw-Hill, 1988.

COHEN, J. C. P.; BELTRÃO, J. C.; GANDU, A. W.; SILVA, R. R. Influence of deforestation on the hydrological cycle in Amazonia. **Ciência e Cultura,** São Paulo, v. 59, n. 3, 2007.

COLLISCHONN, W.; TASSI, R. **Introducing Hydrology.** IPH/UFRGS handout, 2011. Available at: <http://galileu.iph.ufrgs.br/collischonn/apostila_hidrologia/apostila.html>. Accessed on: 22 Dec. 2011.

COSTA, A. C.; DURÃO, R.; PEREIRA, M. J.; SOARES, A. Using stochastic space-time models to map extreme precipitation in Southern Portugal. **Natural Hazards and Earth System Sciences,** v. 8, p. 763-773,2008.

CUTRIM, E. M. C.; MOLION, L. B.; NECHET, D. Rainfall in Amazonia during the 20th Century. In: BRAZILIAN CONGRESS OF METEOROLOGY, 11, Rio de Janeiro, 2000. Proceedings..., Rio de Janeiro: SBMET, 2000.

DAMÉ, R. C. F.; TEIXEIRA, C. F. A.; TERRA, V. S. S. Comparison of different methodologies for estimating Intensity-Duration-Frequency curves for Pelotas - RS. **Engenharia Agrícola,** Jaboticabal, v. 28, n. 2, p. 245-255, 2008.

DAUD, Z. M.; KASSIM, A.H. M.; DESA, M. N. M.; NGUYEN, V. **Statistical analysis of at-site extreme rainfall processes in Peninsular Malaysia.** In: LANEN H A J V, DEMUTH. FRIEND 2002 - Regional Hydrology: Bridging the Gap between Research and Practice.United Kingdom: IAHS Publications, n.2 74, 2002.

KEEPING AN EYE ON THE WEATHER. 23:00 - Heavy rain ALERT for Porto Velho. Available at: <http://wwwdeolhonotempo.blogspot.com/2007/02/2300-alerta-de-chuva-intensa-para-porto.html>. Accessed on: 20 January 2011.

DILL, P. R. J. **Environmental management in river basins.** Santa Maria: UFSM, 2007. Thesis (Doctorate in Agricultural Engineering), Department of Water and Soil Engineering, Federal University of Santa Maria, 2007.

ELIAS, Z.S.; LUIZ ALBERTON, L., VICENTE, E.F.R.; REBELLO, M.; BONIFÁCIO, R.C. Apportionment of indirect costs: application of correlation and regression analysis. **Revista de Contabilidade do Mestrado em Ciências Comerciais da UERJ (online),** Rio de Janeiro, v. 14, n. 2, pp. 50-66, 2009.

ELTZ, F. L.; REICHERT, J. M.; CASSOL, E. A. Rainfall return period in Santa Maria, RS. **Revista Brasileira de Ciência do Solo,** Campinas, v.16, p.265-269, 1992.

FELTRIN, R. M. **Behaviour of the hydrological variables of the soil water balance in drainage lysimeters.** Santa Maria: UFSM, 2009. Dissertation (Master's in Civil Engineering), Department of Water Resources and Environmental Sanitation, Federal University of Santa Maria, 2009.
FIETZ, C. R.; COMUNELL, E. **Probability of rainfall in Mato Grosso do Sul.** Dourados: Embrapa Agropecuária Oeste, 2006.

GAMA, M. J. Clima. In: **Atlas Geoambiental de Rondônia.** 2. ed. Porto Velho: SEDAM, 2003.

GRAPHPAD PRISM. **Graphpad Software Inc.** San Diego: [s.n.], 2007.

GRIMM, A. M.; TEDESCHI, R. G. Influence of El Nino and La Nina events on the frequency of extreme precipitation events in Brazil. In: BRAZILIAN CONGRESS OF METEOROLOGY, 13, 2004, Fortaleza. Proceedings..., Fortaleza: SBMET, 2004.

GUMBEL, E. J. **Statistics of extremes.** New York: Dover Publications, 2004.

HAAN, C. T. **Statistical Methods in Hydrology.** Ames (IA): The Iowa University Press, 1977.

HERNANDEZ, V. Regionalisation of scale parameters in heavy rainfall. **Revista Brasileira de Recursos Hídricos,** v. 13 n. 1, p. 91-98, 2008.

HERSFIELD, D. M.; KOHLER, M. A. An empirical appraisal of the Gumbel extreme value procedure. **Journal of Geophysical Research,** v. 65, n. 6, p. 1737-1746, 1960.

IBGE. Brazilian Institute of Geography and Statistics. **Cities by Federative Units.** Available at: <http://www.ibge.gov.br/cidadesat/topwindow.htm?!>. Accessed on: 20 June 2011.

LEOTTI, V. B.; BIRCK, A. R.; RIBOLDI, J. Comparison of the Kolmolgorov-Smimov, Anderson-Darling, Cramer-Von Mises and Shapiro-Wilk Normality Adherence Tests by simulation. In: SIMPÓSIO DE ESTATÍSTICA APLICADA À EXPERIMENTAÇÃO AGRONÓMICA, 11., 2005, Londrina. **Annals...,** Londrina: SEAEGRO, 2005.

LIMA, F. A. **Bayesian frequency analysis of annual maximum flows with historical information: application to the São Francisco river basin in São Francisco.** Belo Horizonte: UFMG, 2005. Dissertation (Master's Degree in Sanitation, Environment and Water Resources), Department of Hydraulics and Water Resources, Federal University of Minas Gerais, 2005.
MEHL, H. U.; ELTZ, F. L. F.; REICHERT, J. M.; DIDONÉ, I. A. Characterisation of rainfall patterns in Santa Maria (RS). **Revista Brasileira de Ciência do Solo,** Viçosa, v. 25, n. 1,p. 475-483,2001.

MINITAB. **Minitab Statistical Tab.** Pennsylvania: [s.n.], 2011. Available at: <http://www.minitab.com>. Accessed on: 28 November 2011.

MOOG, D. B.; JIRKA, G. H. Analysis of reaeration equations using mean multiplicative error. **Journal of Environmental Engineering,** v. 124, p. 104-110,1998.

MOREIRA, P. S.; DALLACORT, R.; MAGALHÃES, R. A.; INOUE, M. H.; STIELER, M. C.; SILVA, D. J.; MARTINS, J. A. Rainfall distribution and probability of occurrence in the municipality of Nova Maringá-MT. **Revista de Ciências Agro-Ambientais,** Alta Floresta, v. 8, n. 1,p. 9- 20, 2010.

MURTA, R. M.; TEODORO, S. M.; BONOMO, P.; CHAVES, M. A. Monthly rainfall at probability levels using the gamma distribution for two locations in south-west Bahia. **Ciência Agrotécnica,** Lavras, v. 29, n. 5, 2005.

NAGHETTINI, M.; PINTO, E. J. A. **Statistical Hydrology.** Belo Horizonte: CRPM, 2007.

OLIVEIRA, J. P. B.; CECÍLIO, R. C.; XAVIER, A. C.; JASPER, A. P. S.; OLIVEIRA, L. B. Probable rainfall for Alegre-ES using the gamma probability distribution. **Engineering,** v. 7, n. 2, p. 204-211, 2010.

OLIVEIRA, L. F. C.; VIOLA, M. R.; PEREIRA, S.; MORAIS, N. R. Heavy rainfall prediction models for the state of Mato Grosso, Brazil. **Ambi-Agua,** Taubaté, v. 6, n. 3, p. 274-290, 2011.

OLIVEIRA, L. F. C.; CORTÊS, F. C.; BARBOSA, F. O. A.; ROMÃO, P. A.; CARVALHO, D. F. Estimation of heavy rainfall equations for some locations in the state of Goiás using the rainfall disaggregation method. **Pesquisa Agropecuária Tropical,** Goiânia, v. 30, n. 1, p. 23-27, 2000.

PEREIRA, C. E. SILVEIRA, A.; SILVINO, A. N. O. Study of intense rainfall and estimation of the IDF equation for the city of Barra do Bugres - MT. In: SIMPÓSIO DE RECURSOS HÍDRICOS DO CENTRO OESTE, 1., 2007, Cuiabá. Proceedings **...,** Cuiabá: ABRH, 2007.

PEREIRA, S. C. **Sistema de Análise de Risco Operacional: Aplicação de modelos de risco operacional para empresas de produção e serviços não financeiros.** Brasília: UnB, 2010. Monograph (Bachelor's Degree in Control and Automation Engineering), Department of Technology, University of Brasília, 2010.

PRAZERES FILHO, J.; VIOLA, D. N.; FERNANDES, G. B. Use of randomisation test to compare two groups considering non-parametric test. In: SIMPÓSIO NACIONAL DE PROBABILIDADE E ESTATÍSTICA, 19., 2010, São Pedro. Proceedings..., São Pedro: SINAPE, 2010.

PINHEIRO, M. M. G.; NAGUETTI, M. Regional analysis of frequency and temporal distribution of storms in the metropolitan region of Belo Horizonte - RMBH. **Revista Brasileira de Recursos Hídricos,** v. 3, n. 4, p. 73-88, 1998.

PRIOSTE, M. A. O. **Bacia hidrográfica do rio das Ostras: proposta para gestão ambiental sustentável.** Rio de Janeiro, UERJ, 2007. Dissertation (Master's in Environmental Engineering), Department of Environmental Engineering, Rio de Janeiro State University, 2007.

PRUSKI, F. F.; PEREIRA, S. B.; NOVAES, L. F.; SILVA, D. D.; RAMOS, M. M. Average annual precipitation and long-term average specific flow in the São Francisco Basin. **Revista Brasileira de Engenharia Agrícola Ambiental,** Campina Grande, v. 8, n. 2-3, Dec. 2004.

PUYOL, A. F. B.; VILLA, M. A. J. **Princípios y fundamentos de la hidrologia superficial.** Mexico City: Universidad Autónoma Metropolitana, 2006.

QUEIROZ, M. M. F.; SAMPAIO, S. C.; GOMES, B. M.; IOST, C. Study of minimum flows qi,io and q7,io of Paraná rivers according to generalised distribution. **Revista Verde,** Mossoró, v.5, n.3, p. 32-46, 2010.

RAGHUNATH, H. M. **Hydrology principies, analysis and design.** india: New Age International Ltd, 2006.

RELIASOFT - ReliaSoft Corporation. **Characteristics of the Weibull Distribution.** Available at: <http://www.weibull.com/hotwire/issuel4/relbasicsl4.htm>. Accessed on: 03 Dec. 2011.

REYNAUD, F. **Statistical spatial disaggregation of rainfall predicted by the WRF atmospheric model.** Curitiba: UFPR, 2008. Dissertation (Master's in Water Resources and Environmental Engineering), Technology Sector, Federal University of Paraná, 2008.
RIGHETTO, A. M. **Hidrologia e Recursos Hídricos.** São Paulo: EESC/USP, 1998.

RODRIGUES, C.; ADAMI, S. F. Techniques in Hydrography. In: VENTURI, L. A.B.
Geography: Field, Laboratory and Classroom Practices. São Paulo: Ed. Sarandi, 2011.

RONDONIA. State Secretariat for Environmental Development (SEDAM). **Rondônia Climatological Bulletin - 2008.** Porto Velho: SEDAM, 2009.

ROSSA, S. R. L. G. S. **Contributions to a more efficient use of water in the urban cycle: saving water and reusing grey water.** Porto: FEUP, 2006. Dissertation (Master's in Environmental Engineering), FEUP, Faculty of Engineering, University of Porto, 2009.

SÁ, R. L. **Inventory of fluviometric data for the state of Espírito Santo.** Jerônimo Monteiro: UFES, 2011. Monograph (Bachelor's Degree in Forestry Engineering), Department of Forestry Engineering, Federal University of Espírito Santo, 2011.

SANTOS NETO, L. A.; MORAES, J. C.; NOBREGA, R. S. Anthropic Action in the Central Region of Rondônia and the Climate Response. In: BRAZILIAN CONGRESS OF METEOROLOGY, 16, 2010, Belém. **Anais...,** Belém: SBMET, 2010.

SANTOS, I. A.; BUCHMAN, J. A Qualitative Review Emphasising Climatic Aspects of the Amazon and the Northeast Region of Brazil. **Anuário do Instituto de Geociências - UFRJ,** v. 33, n. 2, p. 09-23,2010.

SHERMAN, G. E.; SUTTON, T.; BLAZEK, R.; LUTHMAN, L. **Quantum GIS User Guide - Version 1.7.2,** Seamus, 2011.
SILVA, D. D.; PEREIRA, S. B.; PRUSKI, F. F.; GOMES FILHO, R. R.; LANASE, A. M. Q.; BAENA, L. G. N. Rainfall intensity-duration-frequency equations for the state of Tocantins. **Engenharia na Agricultura,** Viçosa, v. 11, n. 1-4, p.7-14, 2003.

SILVA, D. D.; PINTO, F. R. L.; PRUSKI, F. F.; PINTO, F. A.. Estimation and spatialisation of the parameters of the rainfall intensity-duration-frequency equation for Rio de Janeiro and Espírito Santo. **Engenharia Agrícola,** v. 18, n. 3, p. 22-33. 1999.

SILVA, J. C.; HELDWEIN, A. B.; MARTINS, F. B.; TRENTIN, G.; GRIMM, E. L. Rainfall distribution analysis for Santa Maria, RS. **Brazilian Journal of Agricultural and Environmental Engineering,** v. 11, n. 1, p. 67-72, 2007.

SILVA, L. P.; ZUFFO, C. E. Water resources: conserving for the future. In: **Atlas Geoambiental de Rondônia.** 2. ed. Porto Velho: SEDAM, 2003.

SILVA, M. J. G.; SARAIVA, F. M.; SILVA, A. A. G. Study of Precipitation Behaviour for 2005 in the State of Rondônia. In: BRAZILIAN CONGRESS OF METEOROLOGY, 16, 2010, Belém. **Anais...,** Belém: SBMET, 2010.

SILVA, W. P.; SILVA, C. M. D. P. S.; CAVALCANTI, C. G. B.; SILVA, D. P. S.; SOARES, I. B.; OLIVEIRA, J. A. S. O.; SILVA, C. D. P. S. LAB Fit Ajuste de Curvas: software in Portuguese for processing experimental data. **Revista Brasileira de Ensino de Física,** São Paulo, v.26, n.4, p.419-29, 2004.

SILVINO, A. N. O.; SILVEIRA, A.; MUSIS, C. R.; WYREPKOWSKI, C. C.; CONCEIÇÃO, F. T. Determination of extreme flows for different return periods for the Paraguay River using statistical methods. **Geociências,** São Paulo, UNESP, v. 26, n. 4, p. 369-378, 2007.

TEIXEIRA, W.; FAIRCHILD, T. R.; TOLEDO, M. C. M.; TAIOLI, F. **Decifrando a Terra.** 2. ed. São Paulo: Companhia Editora Nacional - IBEP, v. 1, 2009.

TIBÚRCIO, E. C.; CASTRO, M. A. H. A GIS Application for Determining Average Rainfall Height in River Basins. In: SIMPÓSIO BRASILEIRO DE RECURSOS HÍDRICOS, 17., 2007, São Paulo. Proceedings..., São Paulo: ABRH, 2007.

TORRICO, J. J. T. **Práticas hidrológicas.** Rio de Janeiro: Transcom, 1975.

TUCCI, C. E. M. (Org.). **Hydrology. Science and application.** 4. Ed. Porto Alegre: Ed. da Universidade: ABRH: EDUSP, 2009.
TUCCI, C. E. M. Water in the Urban Environment. In: Aldo da Cunha Rebouças; Benedito Braga; José Galizia Tundisi. (Org.). **Fresh Waters in Brazil.** 1 ed. São Paulo: Escrituras, 1999, v. 1, p. 475-508.

TUCCI, C. E. M. Urban waters. **Estudos Avançados,** v. 22, n. 63, p. 97-112, 2008

TUCCI, C. E. M. **Inundações Urbanas.** 1. ed. Porto Alegre: ABRH, 2007.

VAREJÃO-SILVA, M.A. **Meteorology and climatology.** Brasília: INMET/ Stilo, 2005.

VASCONCELLOS, S. L. B.; ANDRÉ, R. G. B.; PERECIN, D. Sequences of days with and without rain in the municipality of Jaboticabal-SP. **Revista Brasileira de Meteorologia,** v. 14, n. 2, p. 79-90, 1999.

VILLELA, S. M.; MATTOS, A. **Hidrologia aplicada.** São Paulo: McGraw-Hill, 1975.

WEBLER, A. D.; AGUIAR, R. G.; AGUIAR, L. J. G. Rainfall characteristics in primary forest and pasture areas in the state of Rondônia. **Ciência e Natura Magazine,** v. Esp., p. 55-58, 2007.

WERTH, D.; AVISSAR, R. The local and global effects of Amazon deforestation. **Journal of Geophysical Research,** v. 107, n. D20, p. 1-8, 2002.

XAVIER, T. M. B. S.; XAVIER, A. F. S.; DIAS, M. A. F. S.; DIAS, P. L. S. The Intertropical Convergence Zone - ITCZ and its relationship with rainfall in Ceará (1964-98). **Revista Brasileira de Meteorologia,** v. 15, n. 1, p. 27-43, 2000.

XAVIER, T. M. B. S.; XAVIER, A. F. S.; DIAS, M. A. F. S.; DIAS, P. L. S. Interrelations between events (ENSO), the ITCZ in the Atlantic and rainfall in the hydrographic basins of Ceará. **Revista Brasileira de Recursos Hídricos,** v. 8, n. 2, p. 111-126, 2003.

ZOLET, M. **Potential for utilising rainwater for residential use in the urban region of Curitiba.** Curitiba: PUC-PR, 2005. Monograph (Bachelor's Degree in Environmental Engineering), Centre for Exact Sciences and Technology, Pontifical Catholic University of Paraná, 2005.

ZUFFO, C. E.; SILVA, L.P. O Caminho das águas. In: **Atlas Geoambiental de Rondônia.** 2. ed. Porto Velho: SEDAM, 2003.

Printed by Books on Demand GmbH, Norderstedt / Germany